Synthetic Aspects of Biologically Active Cyclic Peptides
——Gramicidin S and Tyrocidines

Contributors

Nobuo IZUMIYA	Laboratory of Biochemistry, Faculty of Science, Kyushu University, Fukuoka 812, Japan
Tetsuo KATO	Laboratory of Biochemistry, Faculty of Science, Kyushu University, Fukuoka 812, Japan
Haruhiko AOYAGI	Laboratory of Biochemistry, Faculty of Science, Kyushu University, Fukuoka 812, Japan
Michinori WAKI	Laboratory of Biochemistry, Faculty of Science, Kyushu University, Fukuoka 812, Japan
Michio KONDO	Department of Chemistry, Faculty of Science and Engineering, Saga University, Saga 840, Japan

Synthetic Aspects of Biologically Active Cyclic Peptides

——Gramicidin S and Tyrocidines

Nobuo IZUMIYA
Tetsuo KATO
Haruhiko AOYAGI
Michinori WAKI
Michio KONDO

A HALSTED PRESS BOOK

KODANSHA LTD.
Tokyo

JOHN WILEY & SONS
New York–Chichester–Brisbane–Toronto

KODANSHA SCIENTIFIC BOOKS

Library of Congress Cataloging in Publication Data

Izumiya, Nobuo.
Synthetic aspects of biologically active cyclic peptides.

(Kodansha scientific books)
"A Halsted Press book."
Includes index.
1. Peptide synthesis. 2. Gramicidins. 3. Tyrocidines. 4. Structure-activity relationship (Pharmacology) I. Title.
QD431.I98 1979 615'.392 79-19998
ISBN 0-470-26863-8

Published in Japan by
KODANSHA LTD.
12–21 Otowa 2-chome, Bunkyo-ku, Tokyo 112, Japan

Published by
HALSTED PRESS
a Division of John Wiley & Sons, Inc.
605 Third Avenue, New York, N.Y. 10016, U.S.A.

PRINTED IN JAPAN

Preface

Recent years have seen dramatic development in the field of peptide chemistry. The importance of conformations of naturally occurring peptides has been emphasized in order to interpret chemical properties and biological activities, just as in protein chemistry. Ever since V. du Vigneaud and his associates verified in 1953 that the chemical synthesis of complex physiologically active peptides is feasible, a good deal of effort has been devoted by organic chemists to the synthesis of naturally occurring peptides and their structural analogs. Hundreds of analogs of biologically active peptides have been synthesized and studied to determine which structural factors might have the biological effect caused by natural molecule.

Most peptide antibiotics from microbial origin have characteristic features and various biological activities. Many of the known peptides are cyclic and contain unusual amino acids, D-amino acids or hydroxy acids in their molecules. Although recent progress in instrumental analysis has led us to make sure of their primary structures and/or conformations in high sensitivity and accuracy, it is still valuable to study synthetic analogs in order to gain much precise information on the structure-function relationships in natural peptides, because structures of the natural peptide antibiotics are often very complexed.

Comprehensive reviews and systematic treatises stressing the organic chemistry of biologically active cyclic peptides such as gramicidin S are still few. In writing this book we attempt to classify a number of microbial peptides into several groups according to their activities and to the type of structure in Chapter 1. Chapter 2 provides a historical summary on isolation and primary structure of gramicidin S and the related peptides such as, tyrocidines, linear gramicidins, gramicidin J and gratisin.

Chapters 3 and 4 focuss on detailed descriptions on chemical aspects of gramicidin S, tyrocidines and their analogs from the viewpoint of their organic synthesis and structure-function relationship. Chapter 5 covers various recent studies on the conformation of the peptide antibiotics by many methods including X-ray, ORD, CD, NMR and IR, and discusses some problems concerning the conformation and the topochemistry of gramicidin S and its analogs. Chapters 6 and 7 deal with mechanism of their antibacterial action and biosynthesis of gramicidin S and tyrocidines. Several topics on organic and biological chemistry of phytotoxic AM-toxins are presented in Chapter 8.

Finally, we should like to acknowledge our co-workers who have discussed, criticized and so helped the development of our ideas over the years. We are also indebted to Miss A. Okabe for typing and Mr. M. Takahatake of Kodansha Scientific who prepared the work for the printer.

Fukuoka
Saga

June, 1979

Nobuo IZUMIYA
Tetsuo KATO
Haruhiko AOYAGI
Michinori WAKI
Michio KONDO

Contents

Abbreviations

The abbreviations suggested for amino acids and peptides (*J. Biol. Chem.*, **247**, 977 (1972)), rules for naming synthetic modifications of natural peptides (*ibid.*, **242**, 555 (1967)) and symbols for the description of the conformation of polypeptide chains (*ibid.*, **245**, 6489 (1970)) are those recommended by the IUPAC–IUB Commission on Biochemical Nomenclature. Unless otherwise specified, the amino acids and peptides are the L stereoisomers. Additional and frequently used abbreviations are as follows:

Ac	Acetyl
AcOEt	Ethyl acetate
Ahp	2-Amino-5-(*p*-hydroxyphenyl)pentanoic acid
Aib	α-Aminoisobutyric acid
βAla	β-Alanine
Amp	2-Amino-5-(*p*-methoxyphenyl)pentanoic acid
Aoe	2-Amino-8-oxo-9,10-epoxydecanoic acid
App	2-Amino-5-phenylpentanoic acid
ATP	Adenosine 5′-triphosphate
δ-Ava	δ-Aminovaleric acid
BBI	Bowman-Birk inhibitor
Boc	*t*-Butyloxycarbonyl
CD	Circular dichroism
C-G	Cytidine-guanosine
Cha	Cyclohexylalanine
Dbu	2,4-Diaminobutyric acid
DCC	Dicyclohexylcarbodiimide
Dha	Dehydroalanine
DMF	Dimethylformamide
DMSO	Dimethylsulfoxide
DNA	Deoxyribonucleic acid
DNP	2,4-Dinitrophenyl
EA	Ethanolamide
EDC	1-Ethyl-3-(3-dimethylaminopropyl)-carbodiimide (water-soluble carbodiimide)
EPR	Electron paramagnetic resonance
GS	Gramicidin S
H_2/Pd	Catalytic hydrogenation
HE	Heavy enzyme

Hmb	2-Hydroxy-3-methylbutanoic acid
Hmp	2-Hydroxy-4-methylpentanoic acid
HOBt	1-Hydroxybenzotriazole
HONSu	*N*-Hydroxysuccinimide
Hyp	Hydroxyproline
IR	Infrared
IE	Intermediate enzyme
LE	Light enzyme
MA	Mixed anhydride
MBzl, MeOBzl	*p*-Methoxybenzyl
mp	Melting point
Ms	Methanesulfonyl
Mz	*p*-Methoxyphenylazobenzyloxycarbonyl
NMR	Nuclear magnetic resonance
NOE	Nuclear Overhauser effect
Nva(4OH)	4-Hydroxynorvaline
OBzl	Benzyl ester
OEt	Ethyl ester
OMe	Methyl ester
ONp	*p*-Nitrophenyl ester
ONSu	*N*-Hydroxysuccinimide ester
ORD	Optical rotatory dispersion
Pht	Phthaloyl
Pip	Pipecolic acid
PMR	Proton magnetic resonance
PPi	Pyrophosphate
Pro(4F)	4-Fluoroproline
RNA	Ribonucleic acid
SPS	Solid-phase synthesis
Suc	Succinyl
TA	Tyrocidine A
TCA	Trichloroacetic acid
TFA	Trifluoroacetic acid
TLC	Thin-layer chromatography
Tos	Tosyl, *p*-toluenesulfonyl
Trt	Trityl, triphenylmethyl
Tyr(Me)	*O*-Methyltyrosine
UV	Ultraviolet
Z	Benzyloxycarbonyl
$Z(NO_2)$	*p*-Nitrobenzyloxycarbonyl
Z(OMe)	*p*-Methoxybenzyloxycarbonyl

CHAPTER **1**

Introduction

Biologically active peptides cover a wide range from dipeptide derivatives, such as penicillin, to polypeptides consisting of over fifty amino acid residues, such as cobrotoxin. Recently, peptides possessing unique activities have been found. For instance, several hormone-releasing hormones and prohormones were obtained from animals, and their structures and functions have been studied. On the other hand, antibiotic or toxic peptides produced by microorganisms are also of interest. Unlike peptides from animals, these microbial peptides contain not only the usual L-amino acids found in proteins but also often D-amino acids, unusual amino acids, hydroxy acids and many other components.

Table 1.1 lists a number of microbial peptides classified into several groups according to their activities. As most studies on natural peptides have been devoted to searches for new substances possessing antimicrobial activity, it is not surprising that many antibacterial, antiyeast or antifungal peptides are included in Table 1.1. Recently, peptides showing unique activities towards insects or plants have been obtained and these are also listed in Table 1.1.

Many other interesting substances have also been found. For example, beauvericin is toxic to the brine shrimp, chlamydocin is cytostatic to mastocytoma cells, and surfactin is a surfactant and has clotting inhibitory activity in the thrombin-fibrinogen reaction. Several proteinase-inhibitory peptides of microbial origin presented in Table 1.1 are very different from those of plant or animal origin in that the microbial peptides generally have a low molecular weight and contain unusual constituents.

On the other hand, the peptides derived from microorganisms can also be grouped by the types of structure, as summarized in Table 1.2. Firstly, they can be divided roughly into linear and cyclic molecules. Secondly, the cyclic peptides can be divided into five types as follows. (1) Low molecular cyclic

TABLE 1.1 Activities and Names of Microbial Peptides

Activity	Name
1) Antibacterial	actinomycins, alamethicin, bacitracin, capreomycins, echinomycin, enniatins, etamycin, gramicidin A, gramicidin S, monamycin, nisin, polymyxins, serratamolide, subtilin, triostins, tyrocidines, valinomycin
2) Antitumor	alazopeptin, albonoursin, duazomycin B, neocarzinostatin
3) Antiyeast, antifungal	amidomycin, mycobacillin, neoantimycin, stendomycin
4) Insecticidal	aspochracin, destruxins
5) Phytotoxic, plant growth inhibitory, chlorosis-inducing	AM-toxins, tabtoxins, Cyl-2, malformin tentoxin
6) Proteinase inhibitory	antipain, bestatin, chymostatin elastinal, leupeptins, pepstatins
7) Iron-transport, siderochrome	enterobactin, ferrichrome

TABLE 1.2 Types and Names of Microbial Peptides

Type	Name
1) Linear	cerexin, gramicidin A, leupeptin, neocarzinostatin, pepstatin
2) Cyclic	
a) Low molecular cyclic peptide	albonoursin, D-cycloserine, penicillin
b) Simple cyclic oligopeptide	antamanide, Cyl-2, fungisporin gramicidin S, malformin, tentoxin, tyrocidines
c) Cyclic oligopeptide having peptidyl side chain	bacitracin, bottromycin, capreomycin, colistin, polymyxins
d) Cyclic peptolide	amidomycin, AM-toxins, destruxins, enniatins, monamycin, serratamolide, sporidesmolide, valinomycin
e) Cyclic peptidelactone	actinomycins, etamycin, stendomycin, vernamycin, viscosin

peptides consisting of amino acid or dipeptide derivatives; the structure of albonoursin (**1.1**) is shown as an example. (2) Simple cyclic oligopeptides containing several amino acids which are linked through a peptide bond;

gramicidin S (**1.2**) is an example. (3) Cyclic oligopeptides having a peptidyl side chain, which can be formed if a peptide of type (2) contains an acidic or basic amino acid residue(s); polymyxin B_1 (**1.3**) is an example. (4) Cyclic peptolides which have one or more hydroxy acid residues in addition to amino acid residues as constituents and therefore have an ester bond in the ring structure; AM-toxin I (**1.4**) and destruxin B (**1.5**) are examples. (5) Cyclic peptidelactones, in which a hydroxyamino acid(s) is present and the hydroxy group instead of the amino group participates in the formation of the backbone structure; viscosin (**1.6**) is an example.

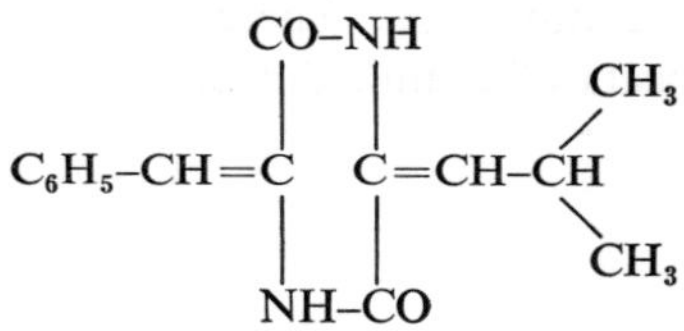

(1.1)

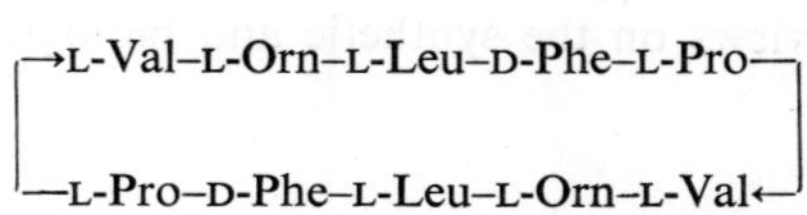

(1.2)

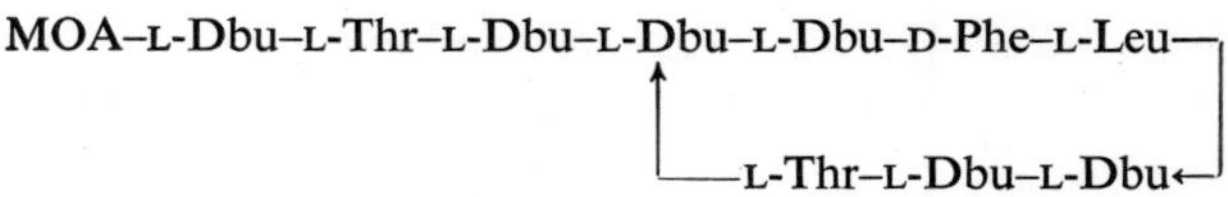

MOA, (+)-6-methyloctanoic acid

(1.3)

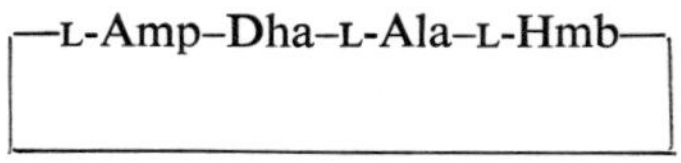

(1.4)

Amp, 2-amino-5-(*p*-methoxyphenyl) pentanoic acid;
Dha, dehydroalanine;
Hmb, 2-hydroxy-3-methylbutanoic acid

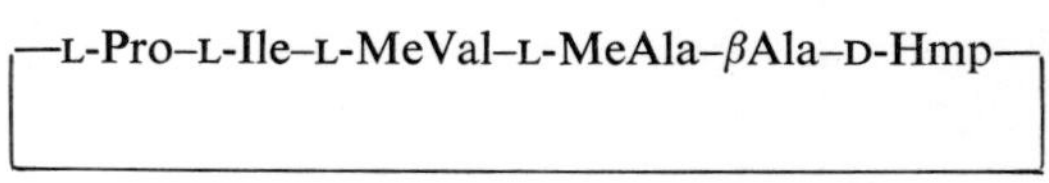

Hmp, 2-hydroxy-4-methylpentanoic acid

(1.5)

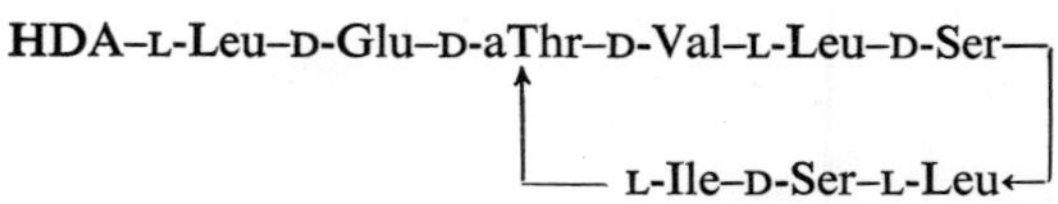

HDA, (−)-3-hydroxydecanoic acid

(1.6)

Among the wide variety of peptide antibiotics, gramicidin S has been one of the most extensively studied by various physico- and organochemical procedures. The reason for this is that its primary structure was determined early in the 1940s, it is available as a pure species from the natural source in fairly amounts, and it has a unique and attractively complex cyclic structure.

In this book, we will describe the chemistry of gramicidin S and related compounds such as tyrocidines, as well as AM-toxins. We have made substantial contributions to the synthetic chemistry of these cyclic peptides, and this book will deal especially with the chemistry of those natural cyclic peptides and analogs synthesized in our laboratory, from the viewpoint of their chemical synthesis and structure-function relationships. There have been reviews on the synthetic and biological aspects of gramicidin S.[1,2]

CHAPTER 2

Isolation and Primary Structures of Gramicidin S and Related Natural Peptides

This chapter deals with the isolation and elucidation of the amino acid sequence of gramicidin S and naturally occurring peptides related to gramicidin S—namely tyrocidines, linear gramicidins, gramicidin J and gratisin.

2.1 Gramicidin S

2.1.1 DISCOVERY AND NOMENCLATURE

A new strain of aerobic sporulating bacteria was isolated from Soviet garden soil by Gause and Brazhnikova in 1944.[3] The isolated bacteria were similar to, though not identical with, a strain isolated by Dubos from *Bacillus brevis* which produced another antibiotic, tyrothricin. The new strain elaborated a hitherto unknown crystalline polypeptide which showed strong antibacterial activity against Gram-positive bacteria. This new substance was designated as gramicidin S; S was taken from Soviet.[4,5]

In 1939, Dubos had already isolated a crude bactericidal agent from cultures of *Bacillus brevis*, which effectively killed Gram-positive microorganisms.[6–9] The isolated crude material was further purified and three crystalline preparations were obtained. Two acid substances, graminic acid and gramidinic acid, were obtained by fractional crystallization of the alcohol solution. The third substance, named gramicidin, was concentrated by repeated recovery of the fraction which remained soluble in alcohol.[10] It was recognized afterwards that this crude preparation, named tyrothricin, was composed of two crystalline components. The first neutral fraction, called gramicidin,[11] contained several linear polypeptide derivatives, and the second material, which was formerly considered to be a mixture of two acid substances called tyrocidines, consisted of a mixture of basic peptides.[12]

The term gramicidin S was proposed by Gause and Brazhnikova, because the material was isolated from a strain of *Bacillus brevis* as a tyrothricin-like substance a few years after Dubos's work. In 1965, Craig[13] pointed out that the term gramicidin S is an unfortunate misnomer and should be altered to gramicidin S-A, since this peptide antibiotic is really a homolog of tyrocidine A (see Section 2.2). However, the term gramicidin S-A has not been adopted by most investigators.

2.1.2 ISOLATION AND CHEMICAL PROPERTIES

The first paper briefly reported that crystalline gramicidin S was obtained in the form of colorless needles with a melting point of 267–268° C by means of alcoholic extraction of the acid precipitate of cultures of *Bacillus brevis*.[5] Subsequently, the method of separation was improved and the yield was increased. For example, Otani *et al.* investigated the optimum conditions for preparation of gramicidin S, and isolated about 0.7 g of gramicidin S per liter of culture of *Bacillus brevis* (Nagano).[14]

The purity of gramicidin S dihydrochloride (as a single molecular species) was gradually improved by the efforts of various investigators. Samples obtained as gramicidin S from culture extracts by different laboratories finally gave identical melting points and specific rotations. Data obtained by the countercurrent distribution technique supported the purity of the samples,[15] and the chemical properties of synthetic gramicidin S agreed well with those of the natural material.[16] The addition of 1 N HCl to ethanol solution of gramicidin S afforded good crystals of columnar form (there is a photograph in the literature[16]) and large crystals were grown by slow cooling of a 70% aqueous ethanol solution. The characteristics of gramicidin S·2HCl reported in the literature are mp 268–270° C,[4] 274–276° C,[17,18] and 277–278° C (decomp.),[16] and specific rotation, $[\alpha]_D$ $-289° \pm 10°$ (70% ethanol),[16] and $-292°$ (70% ethanol).[19] An air-dried sample contains octahydrate ($8H_2O$). The content of water of crystallization depends on the environmental humidity,[18] and a sample vacuum-dried at 80–90° C contains trihydrate ($3H_2O$).[16] The basic gramicidin S (air-dried sample containing $5H_2O$) forms a crystalline powder and shows $[\alpha]_D^{20}$ values of $-314°$ (ethanol) and $-305°$ (methanol) (determined in our laboratory). Synge prepared gramicidin S sulfate, gramicidin S dihydroiodide, gramicidin S flavianate and gramicidin S chloroaurate in order to obtain crystals large enough for X-ray single-crystal analysis. However, he did not report the melting points and specific rotations of these derivatives.[20]

2.1.3 DETERMINATION OF CHEMICAL STRUCTURE

In the year after the discovery of gramicidin S, Synge found five amino acids, L-valine, L-ornithine, L-leucine, D-phenylalanine and L-proline in an acid hydrolysate of gramicidin S, and discovered the absence of carboxyl groups in gramicidin S by electrometric titration. From these findings and the high optical rotation of gramicidin S, he suggested a closed peptide chain structure.[19)]

Consden *et al.*[21)] subjected gramicidin S to partial hydrolysis (10 N HCl–acetic acid, 1:1 by volume) and determined the sequence of the five amino acids in the peptide fragment isolated by ionophoretic fractionation or paper partition chromatography. They also isolated δ-DNP-ornithine from this partial hydrolysate after treatment with 2,4-dinitrofluorobenzene and confirmed Sanger's conclusion[22)] as to the mode of linkage of the ornithine. A sequence of amino acid residues, –Val–Orn–Leu–D-Phe–Pro–, occurring either once or more in a closed peptide chain was proposed for gramicidin S.[23)]

The chemical data on gramicidin S indicated that it must be either a cyclopentapeptide (molecular weight of the free unhydrated base 570) or a cyclodecapeptide (molecular weight of the free unhydrated base 1,140). Pedersen and Synge determined the molecular weight of gramicidin S from the diffusion constant in aqueous acetic acid and ethanol solution and obtained a value in the range of 1,080–1,880, which corresponded to a cyclodecapeptide.[24)] This result was in agreement with the cryoscopic data of Belozerskii and Paskhina for gramicidin S in camphor; the molecular weight as determined in camphor was about 1,060 to 1,340.[25)]

Battersby and Craig separated a reaction mixture of DNP-substituted gramicidin S into mono and diDNP-substituted compounds using the countercurrent distribution method. The molecular weight was calculated from the distribution pattern by weight and by absorption at 350 nm.[15)] These studies provided conclusive proof that gramicidin S is a cyclic decapeptide (**1.2**) as illustrated previously. Subsequent results of X-ray crystallographic studies[26)] and synthetic studies[16)] also supported this cyclic structure of gramicidin S.

2.2 Tyrocidines

Tyrocidine was isolated from tyrothricin, an extract of *Bacillus brevis*, by Hotchkiss and Dubos in 1940.[12)] It was soon shown to consist of basic polypeptides which were fractionated by countercurrent distribution into three major components. The major peptides were called tyrocidine A, tyrocidine

B and tyrocidine C according to their partition ratios in methanol–chloroform –0.1 N HCl (2:2:1 by volume).[27] The characterization of tyrocidine A isolated by recrystallization from methanol and ether gave a melting point of 240–242° C and $[\alpha]_D^{25}$ −111° (c 1.37, 50% ethanol). Determination of its molecular weight (1,270) using countercurrent distribution of DNP-substituted tyrocidine A supported the purity of the antibiotic.[28] Chemical degradation of tyrocidines A, B and C, for example, hydrolysis of tyrocidine A for 6 h in a mixture of acetic acid and conc. HCl at 80° C,[29] similar hydrolysis of tyrocidine B for 3h,[30] and cleavage of tyrocidine C with lithium aluminum hydride,[31] resulted in a single split at the acyl-proline linkage in the molecules. These experiments elucidated the amino acid sequences of the three peptide antibiotics and showed that they are cyclic decapeptides similar to gramicidin S, and differ from each other by a single amino acid replacement (Fig. 2.1).

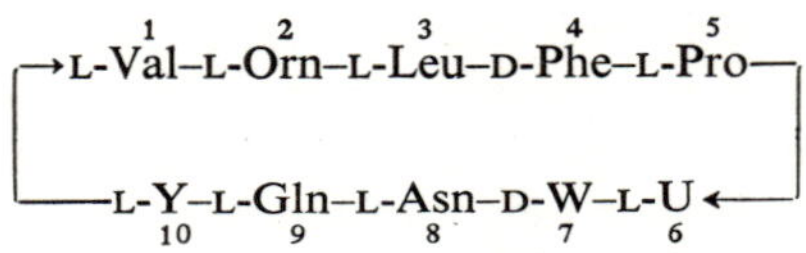

	U	W	Y
Tyrocidine A	Phe	Phe	Tyr
Tyrocidine B	Trp	Phe	Tyr
Tyrocidine C	Trp	Trp	Tyr
Tyrocidine D	Trp	Trp	Trp
Tyrocidine E	Phe	Phe	Phe

Fig. 2.1 Chemical structures of tyrocidine homologs.
(Source: ref. 2. Reproduced by kind permission of Kyoritsu Shuppan Co.)

Ruttenberg and Mach[32] reported that by introducing large amounts of tryptophan into the growth medium of tyrocidines-producing bacteria, it was possible to alter the relative amounts of these three homologs and to induce the incorporation of tryptophan in place of L-tyrosine, thereby producing a hitherto unobserved peptide antibiotic which has been designated as tyrocidine D. Fujikawa *et al.*[33] also reported that the addition of different amino acids to the biosynthetic medium caused a preferential production of new tyrocidines: tyrocidine E (a tyrosine residue in tyrocidine A is replaced by L-phenylalanine), lysyl-tyrocidine A (an ornithine residue in tyrocidine A is replaced by L-lysine) and isoleucyl-tyrocidine A (a leucine residue in tyrocidine A is replaced by L-isoleucine) were synthesized *in vitro* by a cell-free enzyme system of *Bacillus brevis* ATCC 8185.

As can be seen from the structures of the tyrocidines (Fig. 2.1), it is of considerable interest that the pentapeptide sequence, –Val–Orn–Leu–D-Phe–

Pro, in gramicidin S is found in tyrocidines A–E as half of the decapeptide ring. The chemical data presented make it clear that the tyrocidines are very closely related to gramicidin S. However, gramicidin S is produced by one particular strain of *Bacillus brevis* (Nagano, ATCC 9999), while the tyrocidines are produced by Dubos's strain (ATCC 8185). Therefore it is possible that tyrocidine biosynthesis occurs by a mechanism different from that of gramicidin S, especially as regards the specific incorporation of the constituent amino acids.

2.3 Linear Gramicidins

Although the name gramicidin is similar to gramicidin S, gramicidin itself is a linear and neutral peptide antibiotic, and was isolated prior to gramicidin S in 1941 by Hotchkiss and Dubos.[34] They initially assumed a cyclic structure for this substance on the basis that gramicidin contained only neutral amino acids and had no free amino and carboxyl groups.[35] Later, the discovery of ethanolamine in the acid hydrolysate of gramicidin prompted Synge[36] to propose an ortho-peptide structure (**2.1**) for the linkage of ethanolamine to explain the neutral character of this peptide, with an ethanolamide bridge in a cyclic structure.[37]

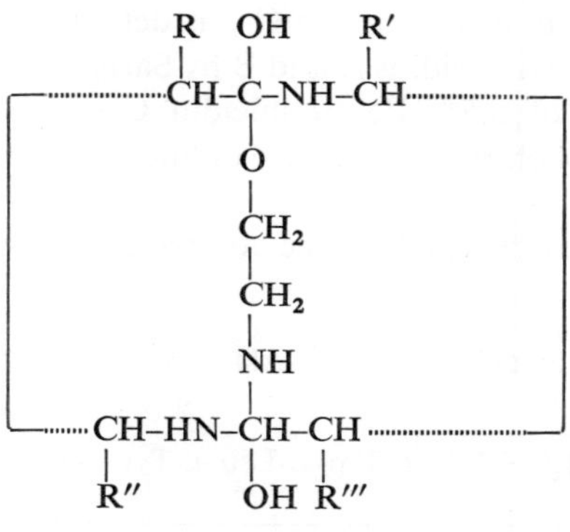

(2.1)

In 1948, Gregory and Craig demonstrated the heterogeneity of natural gramicidin by countercurrent distribution in methanol–water–benzene–chloroform (23:7:15:15 by volume).[38] Craig *et al.* obtained three fractions, of which the major fraction was designated as gramicidin A (distribution coefficient $K = 0.61$); the two minor fractions were designated as gramicidin B ($K = 0.31$) and gramicidin C ($K = 1.32$), respectively.[39,40]

Comparison of quantitative gas chromatograms of the amino acids, which had been converted into DNP-amino acid methyl esters in the hydrolysate before and after the action of D- or L-amino acid oxidase, made possible the complete assignment of configuration to four of the six amino acids present in hydrolysates of gramicidin A.[41]

Treatment of gramicidin A with 1.5 M HCl in methanol at room temperature selectively cleaved one peptide bond and led to the liberation of two amino termini. The sample was found to be a mixture of Val[1]-gramicidin A and Ile[1]-gramicidin A.[42] Sarges and Witkop showed that Val[1]-gramicidin A is a linear pentadecapeptide formylated at the amino terminal residue and masked at the carboxyl terminal as an ethanolamide.[43] The sequences of the two antibiotics were determined by successive Edman degradations and by oxidative cleavage with *N*-bromosuccinimide.[44] In Ile[1]-gramicidin A, the L-valyl residue in position 1 is replaced by a residue of L-isoleucine.[45] The synthesis of these formyl pentadecapeptide ethanolamides and the identification of their properties with those of the natural antibiotics were carried out by Sarges and Witkop.[46]

The primary sequence of gramicidin B, in a sample consisting of 89% Val[1]-gramicidin B and 11% Ile[1]-gramicidin B, was determined by a 14-step Edman degradation of desformylgramicidin B obtained from gramicidin B by mild methanolysis. This sequence very closely resembles that of gramicidin A except for L-phenylalanine in position 11, which replaces L-tryptophan.[47] The chemical structure of gramicidin C was determined in a way analogous to that described for gramicidins A and B by Sarges and Witkop. The sample examined consisted of 5.5% Ile[1]-gramicidin C and 94.5% Val[1]-gramicidin C.[48] The variation of chemical structure of linear gramicidins is shown in Fig. 2.2.

Gross and Witkop[49] resolved the commercially available gramicidin on

$$\overset{1}{\text{HCO–L-X}}\rightarrow\overset{2}{\text{Gly}}\text{–L-}\overset{3}{\text{Ala}}\text{–D-}\overset{4}{\text{Leu}}\text{–L-}\overset{5}{\text{Ala}}\text{–D-}\overset{6}{\text{Val}}\text{–L-}\overset{7}{\text{Val}}\text{–D-}\overset{8}{\text{Val}}\text{–L-}\overset{9}{\text{Trp}}\text{—}$$

$$\text{HOCH}_2\text{CH}_2\text{NH–L-}\underset{15}{\text{Trp}}\text{–D-}\underset{14}{\text{Leu}}\text{–L-}\underset{13}{\text{Tyr}}\text{–D-}\underset{12}{\text{Leu}}\text{–L-}\underset{11}{\text{Y}}\text{–D-}\underset{10}{\text{Leu}}\leftarrow$$

	X	Y
Val-gramicidin A	Val	Trp
Ile-gramicidin A	Ile	Trp
Val-gramicidin B	Val	Phe
Ile-gramicidin B	Ile	Phe
Val-gramicidin C	Val	Tyr
Ile-gramicidin C	Ile	Tyr

Fig. 2.2 Variation in the chemical structures of linear gramicidins.

a preparative scale (60 g) by an extension of Craig's countercurrent distribution method. The distribution pattern indicated the presence of four distinct components, designated as gramicidins A (yield, 27.3 %), B (2.2 %), and C (1.6 %), with a minor component at the far right of the distribution pattern, referred to as gramicidin D (0.1 %). The amino acid sequence of this gramicidin D fraction has not yet been studied in detail. The inhibitory activity on the growth of *Streptococcus faecalis*, however, was found to be slightly stronger with the more hydrophilic C and D fractions (upper phase of benzene–chloroform–methanol–water, 15:15:23:7 by volume) than with the hydrophobic A and B fractions.

Although the primary sequences of gramicidins have been elucidated, some interesting problems remain in connection with their unusual properties and molecular conformation. Gramicidins are generally linear, electrically neutral peptides composed of hydrophobic amino acids with alternating L and D configurations which begin and terminate with formyl and ethanolamide groups, respectively. In order to elucidate the conformation of gramicidin A in solution, several experimental approaches have been utilized. For example, the proton magnetic resonance or circular dichroism studies of Urry *et al.* have suggested that a head-to-head (between the terminal formyl groups) association of two molecules of gramicidin A in a left-handed helix provides a structure of the correct dimensions to span a lipid bilayer membrane and to form a channel for the transfer of ions (see also Chapter 6).[50,51]

Since their organochemical characteristics, such as amino acid composition and primary structure, and physicochemical properties, such as melting point, optical rotation, solubility and molecular conformation, differ from those of gramicidin S, the antibiotic gramicidins have no correlation with gramicidin S except for the name "gramicidin".

2.4 Gramicidin J

In 1958, Otani *et al.*[14] isolated a peptide antibiotic, gramicidin J, from a culture of *Bacillus brevis* (Nagano) and reported it to be a cyclic heptapeptide (**2.2**). Noda[52] later isolated another antibiotic from cultures of *Bacillus brevis* (Nagano) and determined its chemical structure (**2.3**).

cyclo(–L-Val–L-Orn–D-Phe–D-Leu–L-Phe–L-Pro–D-Orn–)

(2.2)

cyclo(–L-Val–D-Orn–L-Pro–D-Phe–D-Leu–L-Orn–)

(2.3)

Thus, the former was named gramicidin J_1 (**2.2**) and the latter gramicidin J_2 (**2.3**). Noda suggested that the original strain of *Bacillus brevis* (Nagano) which produced J_1 (**2.2**) might have mutated to one producing J_2 (**2.3**). He reported a chemical synthesis of a cyclic hexapeptide corresponding to J_2 (**2.3**), which was proposed to be identical with the natural product. The synthesis of J_2 was repeated by Kato and Izumiya.[53)] In a comparison of various properties, such as biological activity, $[\alpha]_D$ and melting point between the synthesized cyclohexapeptide corresponding to **2.3** and the natural J_2 reported by Noda, serious differences were observed.[54)] During the course of Kato's synthetic studies, Kurahashi[55)] and Otani *et al.*[56)] reinvestigated the chemical structure of natural J_1 and J_2, and found that both of them were identical to gramicidin S. Otani *et al.* withdrew the chemical structure proposed for gramicidin J_1. However, they carried out unique investigations on the mechanism of antibacterial action of natural gramicidin J_1 from around 1955 to 1960; their results are significant for an understanding of the mechanism of antibacterial activity of gramicidin S.

2.5 Gratisin

An antibiotic, gratisin, was isolated from the cells of a mutant strain, *Bacillus brevis* Y-33, by Zharikova *et al.*[57)] The antibiotic basic peptide was extracted with ethanol from the dry cells and purified by carboxymethylcellulose column chromatography using a buffer system of pyridine–acetic acid–ethanol–water at pH 5.1. Recrystallization from aqueous ethanol–chloroform–petroleum ether gave colorless crystals, mp 274–276° C, with a characteristic UV absorption maximum in 0.1 N HCl at 270 nm. This compound is soluble in methanol, ethanol, acetic acid and aqueous acetone, poorly soluble in water, and insoluble in chloroform, benzene and ether. Comparative chromatography and electrophoresis of gratisin and gramicidin S in various solvent systems indicated a slight difference in mobility between the two antibiotics.[57)]

The structure of gratisin was determined to be a cyclic dodecapeptide

→Val–Orn–Leu–Phe–Pro–Tyr–

–Tyr–Pro–Phe–Leu–Orn–Val←

(2.4)

(**2.4**) by distribution chromatography and by amino acid sequence analysis of the acid hydrolysate. In addition electrophoretic studies revealed that the

DNP-derivative of gratisin consists of a mixture of mono and diDNP-gratisin which contain δ-DNP-ornithine and no free α-amino group, respectively. Myaskovskaya *et al.* did not define the L or D configuration of the constituent amino acids. The value of molecular weight determined by Battersby and Craig's method using DNP-gratisin was in good agreement with the theoretical value for the cyclic structure (**2.4**).[58] At present, no further information on biological studies or the chemical synthesis of gratisin is available.

CHAPTER **3**

Chemical Synthesis of Gramicidin S and Related Peptides

Synthetic methods for structurally complicated linear oligopeptides were already well-developed by 1955. For example, the pituitary hormone oxytocin was successfully synthesized by du Vigneaud in 1953.[59]

However, virtually no attempts to synthesize cyclic peptides were made before 1950. Since the early 1950's, the number of studies describing syntheses of cyclic peptides has steadily increased, and this reflects the growing interest of organic chemists in the chemical synthesis of peptide antibiotics having unusual cyclic structures. Synthetic studies on cyclic peptides have been undertaken mainly to confirm or test proposed structures and for the preparation of analogs with biological activity. In 1952, Boissonnas[60] reported the synthesis of cyclo(–D-Leu–Gly–Gly–)$_2$, a cyclic peptide, by the use of modern techniques of peptide chemistry. The developments in the synthesis of cyclic peptides were reviewed by Kato *et al.*[61] in 1966 and Kopple[62] in 1972. At present, further fundamental studies on the synthesis of cyclic peptides are still required. For instance, Rothe *et al.*[63] recently studied the tendency of peptides containing proline and sarcosine to cyclize under appropriate conditions.

Amino acids with a third functional group in addition to the amino and carboxyl groups require additional consideration in the selection of protecting groups and of coupling methods. Schwyzer *et al.* developed not only new methods for cyclization but also new ideas on suitable combinations of protecting groups. Schwyzer and Sieber succeeded in the chemical synthesis of gramicidin S, thus synthesizing for the first time a naturally occurring cyclic peptide.[16] Since then many methods aiming at a better yield in the cyclization reaction have been widely investigated using gramicidin S or gramicidin S analogs as model compounds. Gramicidin S seems to occupy a key position among cyclic peptides. We will next describe reported procedures for the chemical synthesis of gramicidin S and related peptides.

3.1 Synthesis of Gramicidin S

Cyclic peptides are usually prepared from linear peptides by intramolecular cyclization. The carboxylic end of the peptide is usually activated by the formation of active esters, anhydrides, azides, or chlorides, and the free terminal amino end is allowed to react wtih the active terminal carboxyl group at high dilution. There are two main problems in cyclic peptide synthesis: first, construction of the open chain peptide, and second, the cyclization (ring formation) reaction.

In the case of gramicidin S, five linear decapeptides can theoretically be used for the cyclization: arrows in structure **3.1** indicate possible combinations of *N*- and *C*-terminal amino acid residues in open chain decapeptides. However, activation of a proline residue at the *C*-terminus has been widely

$$\begin{array}{c} \rightarrow \text{L-Val} \cdot\overset{\downarrow}{\cdot}\cdot \text{L-Orn}(\delta\text{-R}) \cdot\overset{\downarrow}{\cdot}\cdot \text{L-Leu} \cdot\overset{\downarrow}{\cdot}\cdot \text{D-Phe} \cdot\overset{\downarrow}{\cdot}\cdot \text{L-Pro} \cdot\overset{\downarrow}{\cdot}\cdot \text{L-Val} \text{—}\urcorner \\ \llcorner\text{———— L-Pro–D-Phe–L-Leu–L-Orn}(\delta\text{-R}) \leftarrow \text{————}\lrcorner \end{array}$$

(3.1)

employed in order to avoid racemization. Recently a new approach to ring formation from the linear peptide containing *C*-terminal L-leucine has attracted marked interest because this approach simulates the biosynthetic route, in which the peptide chain is elongated by enzymatic reaction from the *N*-terminal D-phenylalanine to construct gramicidin S.

In the following discussion, we will divide the synthetic methods for gramicidin S into two classes; (1) solution methods and (2) solid-phase methods.

3.1.1 SOLUTION METHODS

Among the methods that allow cyclization in solution, the *p*-nitrophenyl ester, *N*-hydroxysuccinimide ester, azide and phosphite methods have been most widely employed in the preparation of gramicidin S.

A. The *p*-Nitrophenyl Ester Method

This active ester method involves the cyclization of a linear decapeptide active ester or the cyclodimerization of a linear pentapeptide active ester corresponding to the gramicidin S sequence.

a. Cyclization of a Linear Decapeptide

Schwyzer and Sieber achieved the synthesis of gramicidin S in 1957.[16] This was the first chemical preparation of a macrocyclic peptide antibiotic. The synthetic routes are outlined in Fig. 3.1.

Fig. 3.1 Total synthesis of gramicidin **S**.

Z–L-Val–L-Orn(Tos)–OMe (**3.2**) and Z–L-Leu–D-Phe–L-Pro–OMe (**3.3**) were obtained by the active ester method. Conversion of **3.2** to the hydrazide and catalytic hydrogenation of **3.3** were followed by condensation to form Z–L-Val–L-Orn(Tos)–L-Leu–D-Phe–L-Pro–OMe (**3.4**). The benzyloxycarbonyl (Z) group was removed by hydrogenation and then a trityl (Trt) group was introduced. Alkaline hydrolysis yielded the acylpentapeptide acid (**3.6**), which was coupled using 1-cyclohexyl-3-(2-morpholinyl-(4)-ethyl)-carbodiimide with the pentapeptide ester (**3.5**). After saponification of **3.7**, the acyldecapeptide acid (**3.8**) was converted to the active ester (**3.9**) by the action of di-*p*-nitrophenyl sulfite. The *p*-nitrophenyl ester content of the resulting **3.9** was estimated to be 92% by comparison of the optical density at 400 nm with that of standard Z–Leu–ONp. Detritylation with trifluoroacetic acid and then lyophilization yielded the decapeptide *p*-nitrophenyl ester trifluoroacetate. The cyclization reaction was carried out by dropwise addition of a solution of the active ester trifluoraocetate in DMF to a large amount of pyridine at 55° C; the concentration of the decapeptide *p*-nitrophenyl ester in pyridine was 2.8×10^{-3} M. After the solvent had been removed, the crude product was purified by passing it through columns of acidic and basic ion exchangers to remove by-products and it was subsequently purified by countercurrent distribution and by neutral alumina column chromatography; the yield of crystalline cyclodecapeptide (**3.10**) from **3.8** was 28% after recrystallization from aqueous ethanol. This **3.10** was identical with the tosylated product of natural gramicidin S. The tosyl (Tos) groups of **3.10** were removed by reduction with sodium in liquid ammonia. The resulting cyclo(–L-Val–L-Orn–L-Leu–D-Phe–L-Pro–)$_2$·2HCl was purified by countercurrent distribution and crystallized from ethanol and 1 N HCl. The synthetic gramicidin S thus obtained was found to be identical with natural gramicidin S·2HCl on the basis of a comparison of melting point, specific rotation, R_f values on paper chromatograms in various solvent systems, the infrared spectrum, X-ray diffraction patterns and microbiological assays.

This work was noteworthy in the following respects. (1) Planning of the synthetic route for the linear peptide and the cyclization, (2) introduction of a new reagent (di-*p*-nitrophenyl sulfite) for forming active esters, and (3) the combination of temporary and "permanent" protecting groups.

b. Cyclization of a Linear Pentapeptide

In 1958, Schwyzer and Sieber[64)] reported further interesting work on the cyclization of a linear pentapeptide active ester (**3.11**). They found that instead of the expected monomeric cyclopentapeptide, cyclization of **3.11** led to diTos-gramicidin S (**3.10**) by condensation of two molecules of **3.11**.

$$\underset{(\mathbf{3.11})}{\text{H–L-Val–L-Orn(Tos)–L-Leu–D-Phe–L-Pro–ONp}} \xrightarrow[\text{in pyridine}]{27\%\ \text{yield}} \underset{(\mathbf{3.10})}{\text{diTos-gramicidin S}}$$

Schwyzer[65] explained the mode of this dimeric cyclization as follows. When the tripeptide sequences –L-Val–L-Orn–L-Leu– of the two molecules associate by hydrogen bonding and take the antiparallel β-pleated sheet structure proposed by Pauling and Corey,[66] the *C*-terminal active esters (X-position in Fig. 3.2) will each be placed opposite the amino group of another molecule because the peptide backbone kinks at the positions of D-phenylalanine and L-proline residues, as shown in Fig. 3.2. This orientation will lead to a dimeric cyclization reaction and affords a favorable conformation for gramicidin S, as depicted in Fig. 3.3. Schwyzer also mentioned that the phenomenon might be expected to occur especially effectively with peptides containing an odd number of amino acid residues.

In the initial study in our laboratory on the synthesis of a gramicidin S

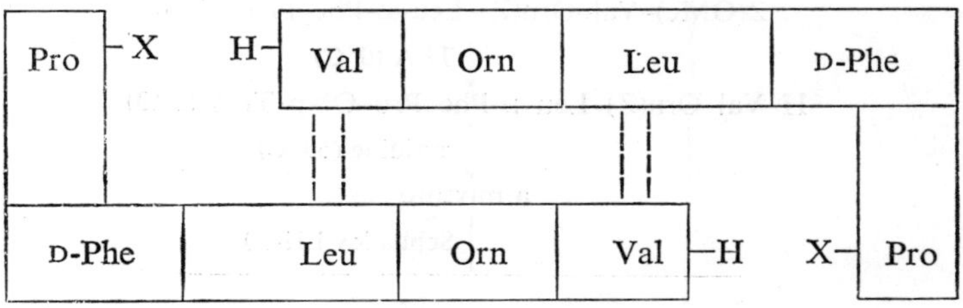

Fig. 3.2 Possible mode of doubling cyclization.
(Source: ref. 2. Reproduced by kind permission of Kyoritsu Shuppan Co.)

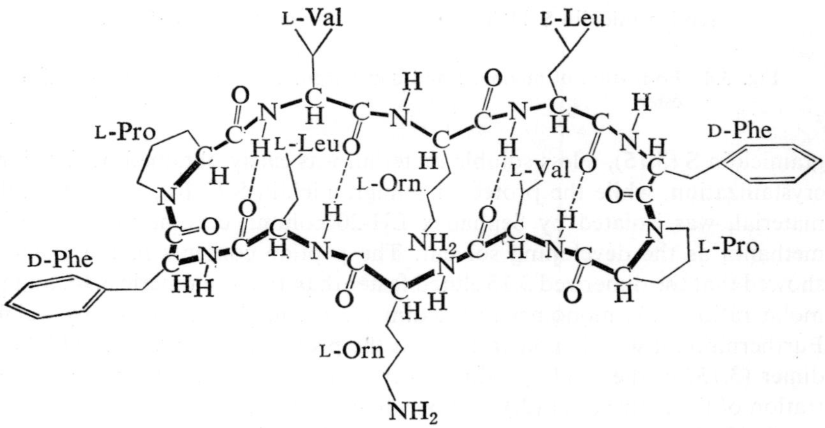

Fig. 3.3 A model of gramicidin S with β-pleated sheet structure.

analog, Aoyagi *et al.*[67] intended to obtain a dimerized product by the cyclization of a linear pentapeptide active ester (**3.12**) according to Schwyzer's prediction. However, the main product was confirmed to be a monomeric

H–L-Val–L-Orn(Z)–L-Leu–D-Phe–Gly–ONp

(**3.12**)

cyclopentapeptide by means of paper electrophoresis and molecular weight determination with an osmometer. These results prompted Waki and Izumiya[17,18] to investigate the possibility of obtaining semigramicidin S. It appeared that the crude product of cyclization of the linear pentapeptide active ester (**3.13**) was composed of two components (Fig. 3.4). The protected

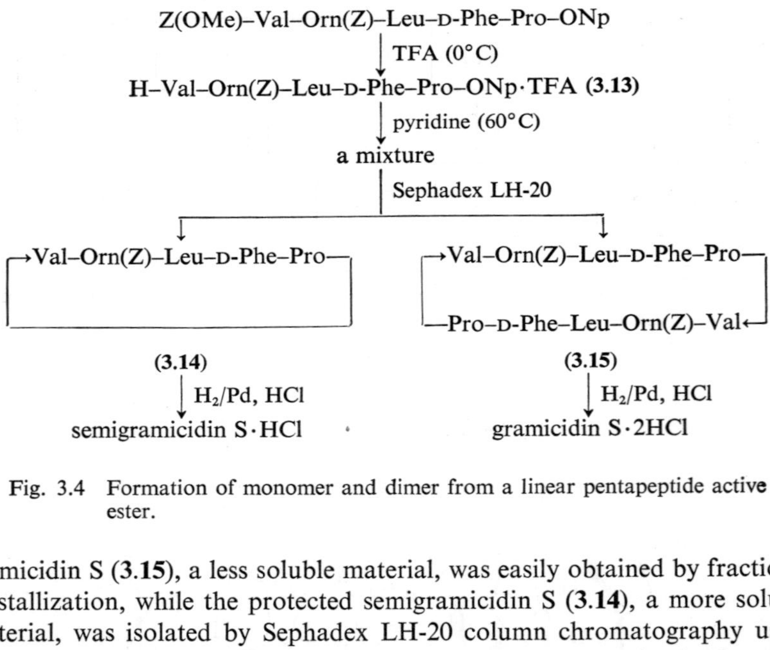

Fig. 3.4 Formation of monomer and dimer from a linear pentapeptide active ester.

gramicidin S (**3.15**), a less soluble material, was easily obtained by fractional crystallization, while the protected semigramicidin S (**3.14**), a more soluble material, was isolated by Sephadex LH-20 column chromatography using methanol as the developing solvent. The peptide contents in the fractions showed that the dimerized **3.15** eluted faster than the monomeric **3.14**, and the molar ratio of the monomer to the dimer was found to be 3 to 7 by weight. Furthermore, it was demonstrated that the proportion of monomer (**3.14**) to dimer (**3.15**) in the crude product after cyclization depends on the concentration of the active ester (**3.13**) in pyridine solution.

Table 3.1 shows that a very dilute solution of pentapeptide active ester (**3.13**) is favorable to the monomeric cyclization, whereas a strong intermole-

TABLE 3.1 Effect of the Concentration of a Pentapeptide Active Ester (**3.13**) on the Cyclization Yields of Monomer and Dimer
(Source: ref. 2. Reproduced by kind permission of Kyoritsu Shuppan Co.)

Concentration of **3.13** in pyridine	Weight ratio of cyclic peptides	
(M)	Monomer (**3.14**)	Dimer (**3.15**)
0.3×10^{-3}	45	55
3×10^{-3}	32	68
30×10^{-3}	29	71

cular association of the peptide (**3.13**) causing dimerization is still present in a highly dilute solution in pyridine, even at concentrations as low as 0.3 mM of **3.13**.

B. The *N*-Hydroxysuccinimide Ester (ONSu) Method and the Azide Method

The cyclization yields of the linear peptide *p*-nitrophenyl ester described above were 30–40%, which were not sufficient for usual peptide synthesis. In order to search for a more effective procedure to give a better yield than the ONp method, the formation of activated linear pentapeptide (**3.16**) or decapeptide (**3.17**) by the ONSu method or the azide method was investigated in our laboratory.[68)]

H–L-Val–L-Orn(Z)–L-Leu–D-Phe–L-Pro–X

(**3.16**)

H–(L-Val–L-Orn(Z)–L-Leu–D-Phe–L-Pro)$_2$–X

(**3.17**)

In the cyclization of **3.16**, Boc-pentapeptide *N*-hydroxysuccinimide ester was prepared from Boc–L-Val–L-Orn(Z)–L-Leu–D-Phe–L-Pro–OH (**3.18**) by the action of HONSu and water-soluble carbodiimide (EDC), and its Boc group was removed with trifluoroacetic acid. The pentapeptide ester trifluoroacetate (**3.16**; X = ONSu) thus obtained was treated with pyridine at a concentration of 3×10^{-3} M of **3.16** for cyclization at room temperature to afford a mixture of Z-substituted cyclic pentapeptide (**3.14**) and decapeptide (**3.15**). The mixture was subjected to hydrogenolysis. Separation of the components was achieved with a column of carboxymethylcellulose using a linear salt gradient, and each product obtained was converted to its hydrochloride to yield pure crystals of cyclic pentapeptide hydrochloride (semigramicidin S·HCl) or decapeptide dihydrochloride (gramicidin S·2HCl). On the other hand, Boc–L-Val–L-Orn(Z)–L-Leu–D-Phe–L-Pro–N_2H_3 was also converted to a pentapeptide hydrazide dihydrochloride with hydrogen chloride in formic acid[69] and then transformed into a pentapeptide azide with isoamyl nitrite. The azide was treated with pyridine at a concentration of 3×10^{-3} M at 4° C for 3 days to yield a mixture of monomer (**3.14**) and dimer (**3.15**). The mixture was worked up by the procedure described for the ONSu method. The cyclization of the activated linear decapeptide (**3.17**; X = ONSu or N_3) was achieved by the same procedure used in the cyclization of **3.16** and yielded exclusively a protected cyclic decapeptide (**3.15**) which was hydrogenated to afford gramicidin·2HCl (Fig. 3.5).

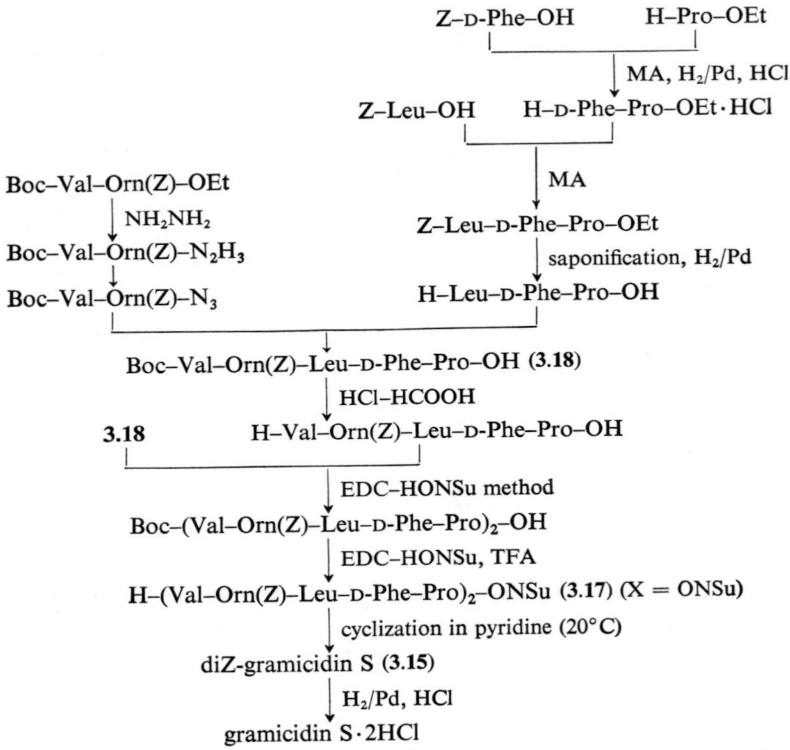

Fig. 3.5 Synthesis of gramicidin S by the *N*-hydroxysuccinimide ester method.

The results summarized in Table 3.2 indicate that the ONSu method is more effective for cyclization than the azide or ONp method with respect to good yield, mild experimental conditions and simple procedure at the cyclization step.

A new approach was developed by Tanaka *et al.* to prepare another acti-

TABLE 3.2 Yields of Semigramicidin S (semiGS) and Gramicidin S (GS) from Activated Pentapeptide (**3.16**) and Decapeptide (**3.17**) by Three Methods (Source: ref. 2. Reproduced by kind permission of Kyoritsu Shuppan Co.)

Method	Yield from pentapeptide (**3.16**)			Yield of diZ-GS from decapeptide (**3.17**)
	Crude cyclic peptide	semiGS·HCl	GS·2HCl	
ONSu	60%	14%	38%	72%
Azide	50%	12%	30%	72%
ONp	40%	14%	20%	28%

vated linear pentapeptide in 1967.[70] The basic concept was to simulate the biosynthetic pathway of gramicidin S by *Bacillus brevis*. As described in Chapter 7, a pentapeptide fragment of gramicidin S containing L-leucine at the *C*-terminus is dimerized on the surface of an enzyme in the bacteria. A linear pentapeptide azide (**3.19**) was subjected to cyclization by the azide method, which minimizes racemization of *C*-terminal L-leucine. The pure diZ-gramicidin S (**3.15**) was obtained in a yield of 30–36% from **3.19**. However, Tanaka *et al.* did not mention whether or not a cyclic monomer was formed.

$$\underset{\textbf{(3.19)}}{\text{H–D-Phe–L-Pro–L-Val–L-Orn(Z)–L-Leu–N}_3} \xrightarrow[\text{in pyridine}]{30\text{–}36\%\ \text{yield}} \underset{\textbf{(3.15)}}{\text{diZ-gramicidin S}}$$

Nonaka *et al.*[71] studied the cyclization of pentapeptides with the peptide H–D-Phe–L-Pro–L-Val–L-Orn(Z)–L-Leu–X (X = ONSu or N_3), in which L-leucine occupies the *C*-terminus, and examined how variation of activation methods and of reaction concentrations influenced the composition of the product. Table 3.3 shows that cyclization of these pentapeptides yielded a

TABLE 3.3 Ratio of Monomer to Dimer after Cyclization of the Pentapeptide H–D-Phe–L-Pro–L-Val–L-Orn(Z)–L-Leu–X†
(Source: ref. 71. Reproduced by kind permission of Protein Research Foundation)

X	Concentration (mM)	Monomer	:	Dimer	Yield (%)
ONSu	30	44	:	56	40
	3	57	:	43	60
	0.3	71	:	29	87
N_3	30	60	:	40	27
	3	75	:	25	45
	0.3	85	:	15	67

† Reaction conditions; ONSu, at room temperature for 2 h; azide, at 0°C for 2 h and then 5°C for 3 days.

large amount of cyclic monomer. The contents of dimer in the reaction products were lower than in those prepared from precursors with *C*-terminal L-proline. This result differed from that of Tanaka *et al.* and suggested that the mode of biosynthetic cyclization may differ from that of chemical synthesis.

C. The Phosphite Method

Rothe *et al.*[63] used a phosphite method for the cyclization of a pentapeptide (**3.20**). In this method, two reaction mechanisms can be considered, one *via* activation of an α-amino group in the linear peptide and the other *via* an activated carboxyl terminus in the peptide. The *C*-terminal carboxyl group, which is activated in the presence of phosphite reagent by a mixed anhydride, reacts spontaneously with a free amino group to form a cyclic peptide (**3.21**) (Fig. 3.6). The main advantages of this method are the experi-

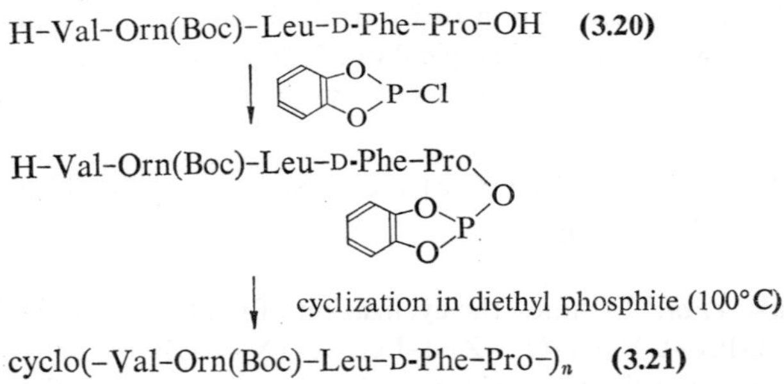

Fig. 3.6 Synthesis of gramicidin S by the phosphite method.

mental simplicity and high yield (up to 80%). The total yield of the cyclized product (**3.21**) was 87%, and the product was found to be a mixture of monomer and dimer in the ratio of 62 to 38. Rothe *et al.* suggested that this can be explained by a stronger association occurring in pyridine than in diethyl phosphite used in the phosphite method. In accord with this view, in the latter solvent they found a reversed yield ratio of both cyclic peptides at concentrations of 5×10^{-3} M.

3.1.2 THE SOLID-PHASE METHOD

Most syntheses of gramicidin S have been carried out by solution methods. A few examples of the synthesis of gramicidin S using solid supports have been reported, however.

A. Cyclization with Dicyclohexylcarbodiimide

Klostermeyer[72)] prepared gramicidin S by a solid-phase method using Merrifield resin (Fig. 3.7). The synthesis of a linear decapeptide-resin (**3.22**) was achieved by two different procedures. One involved the use of dicyclohexylcarbodiimide as the coupling reagent with Boc-amino acids, and the other employed an active ester (ONSu) of Boc-amino acids. After cleavage from the supporting polymer resin, a pure linear decapeptide (**3.23**) was obtained only by the former procedure. The cyclization of **3.23** with an excess of DCC and *p*-nitrophenol produced the desired ditosyl cyclodecapeptide in pure form in 45% yield. This diTos-gramicidin S was identical with the authentic sample made by Schwyzer and Sieber. After removal of tosyl groups with sodium in liquid ammonia, the synthetic gramicidin S·2HCl·$3H_2O$ was recrystallized from ethanol and 1 N HCl; mp 281–282° C and $[\alpha]_D - 283° \pm 5°$ (70% ethanol).

Boc–Pro–resin

↓ stepwise addition of Boc–amino acids

Boc–(Val–Orn(Tos)–Leu–D-Phe–Pro)$_2$–resin (**3.22**)

↓ HBr–TFA

H–(Val–Orn(Tos)–Leu–D-Phe–Pro)$_2$–OH (**3.23**)

↓ DCC, *p*-nitrophenol

diTos-gramicidin S (**3.10**)

Fig. 3.7 Solid-phase synthesis of gramicidin S *via* the decapeptide.

A combination of the furfuryloxycarbonyl group for α-amino protection and a resitol-type resin was used in another solid-phase synthesis of gramicidin S.[73)] The yield and biological activity of the final product were similar to those obtained by the usual solid-phase method.

B. The Azide Method and Fragment Solid-Phase Method

Ohno *et al.*[74)] has reported an improved synthesis of gramicidin S by a solid-phase method. An outline of the synthesis is given in Fig. 3.8. The protected linear decapeptide sequence of gramicidin S was built up by stepwise addition of *p*-methoxybenzyloxycarbonyl amino acids to the Merrifield resin, starting with L-leucine as the *C*-terminal amino acid. Hydrazinolytic cleavage of the protected decapeptide hydrazide (**3.24**) from the resin support was carried out by the action of excess hydrazine hydrate. The resulting decapeptide hydrazide (**3.24**) was estimated to be 98% pure on the basis of its absorption at 268 nm due to the *p*-nitrobenzyloxycarbonyl and *p*-methoxybenzyloxycarbonyl chromophores involved. Cyclization by the azide method in

Z(OMe)–Leu–resin

↓ stepwise addition of Z(OMe)-amino acids

Z(OMe)–(D-Phe–Pro–Val–Orn(Z(NO_2))–Leu)$_2$–resin

↓ $NH_2NH_2 \cdot H_2O$ in DMF

Z(OMe)–(D-Phe–Pro–Val–Orn(Z(NO_2))–Leu)$_2$–$NHNH_2$ **(3.24)**

↓ HCl in ethyl acetate, anisole

H–(D-Phe–Pro–Val–Orn(Z(NO_2))–Leu)$_2$–$NHNH_2 \cdot 2HCl$

↓ HNO_2

H–(D-Phe–Pro–Val–Orn(Z(NO_2))–Leu)$_2$–$N_3 \cdot HCl$ **(3.25)**

↓ pyridine

cyclo(–D-Phe–Pro–Val–Orn(Z(NO_2))–Leu–)$_2$ **(3.26)**

↓ H_2/Pd, HCl

gramicidin S·2HCl

Fig. 3.8 An improved solid-phase synthesis of gramicidin S by the azide method.

pyridine furnished a protected cyclodecapeptide (**3.26**) in a good yield (85%). The hydrogenolyzed product exhibited the same activity as natural gramicidin S.

The solid-phase method has obvious merits as regards simplifying and accelerating peptide synthesis. However, one serious problem in this method is the unavoidable accumulation of contaminating by-products, and this makes purification of the desired products very difficult. Thus, the application of a fragment condensation procedure to the solid-phase method is undoubtedly an attractive approach for the improvement of the method.[75)]

In order to examine the usefulness of the fragment condensation procedure and to compare the stepwise elongation procedure and the fragment condensation procedure in the solid-phase method, the synthesis of a linear decapeptide corresponding to the sequence of gramicidin S was studied by Sato *et al.*[76)] using three procedures as follows. (1) Procedure a, stepwise elongation using Boc-amino acids. (2) Procedure b, fragment condensation with Boc-dipeptide fragments except for Boc-D-phenylalanine at the *N*-terminus. (3) Procedure c, fragment condensation with Boc-tripeptide fragments (Fig. 3.9). The purity of each synthetic intermediate peptide was ascertained spectrophotometrically by measuring the absorption of the Z(NO_2) group which protects the δ-amino group of L-ornithine. Hydrazinolysis of the peptidyl resin (**3.27**) was carried out as described above, and cyclization of **3.25** by the azide method gave a protected cyclic decapeptide (**3.26**). The gramicidin S thus synthesized by the three procedures exhibited the same chemical properties and the same antibacterial activities as natural gramicidin S. These results did not suggest any advantage of this fragment solid-phase

H–Leu–resin

↓ a, Boc-amino acid + DCC
b, Boc-dipeptide acid + HOBt + DCC
c, Boc-tripeptide acid + HOBt + DCC

Boc–(D-Phe–Pro–Val–Orn(Z(NO_2))–Leu)$_2$–resin (**3.27**)

↓ NH_2NH_2, 0.1 N HCl–HCOOH

H–(D-Phe–Pro–Val–Orn(Z(NO_2))–Leu)$_2$–$NHNH_2 \cdot 2HCl$ (**3.25**)

↓ HNO_2, pyridine

cyclo(–D-Phe–Pro–Val–Orn(Z(NO_2))–Leu–)$_2$ (**3.26**)

↓ H_2/Pd, HCl

gramicidin S·2HCl

Fig. 3.9 Synthesis of gramicidin S by the fragment solid-phase method.

method other than slightly greater ease of purification. However, this can be important when one constructs longer chain peptides than gramicidin S.

3.2 Synthesis of Analogs of Gramicidin S

Typical analogs of gramicidin S contain amino acid substitutions formed by replacement of one or more of the constituent amino acids of gramicidin S; modified compounds can also be obtained which are derivatives of natural gramicidin S. Furthermore, linear or cyclic peptides with a partial sequence of gramicidin S and with smaller or larger molecular size than gramicidin S will also be discussed here as analogs of gramicidin S. The synthetic strategy for amino acid replacement analogs is essentially the same as that described for the synthesis of gramicidin S in the preceding section. A wide variety of amino acid replacement analogs can be prepared by stepwise chemical synthesis, and these can be used to study the structure-activity relationships of gramicidin S in more detail. A modified analog can easily be obtained by chemical modification of natural gramicidin S.

In this section our attention will be focused on cyclization methods used for the synthesis of analogs of gramicidin S. The following numbering for amino acid residues in gramicidin S will be used. Thus, for example, the ana-

(1) (2) (3) (4) (5)
→L-Val–L-Orn–L-Leu–D-Phe–L-Pro—┐
└—L-Pro–D-Phe–L-Leu–L-Orn–L-Val←┘
(5′) (4′) (3′) (2′) (1′)

Gramicidin S

log, in which L-proline residues at the 5 and 5′ positions in gramicidin S are replaced by glycine residues is expressed as [Gly$^{5,5'}$]-gramicidin S.

3.2.1 VARIOUS CYCLIZATION METHODS

The following cyclization methods will be described here: the *p*-nitrophenyl ester, *N*-hydroxysuccinimide ester, azide, phosphite, dicyclohexylcarbodiimide–*N*-hydroxysuccinimide and acid chloride methods, and cyclization on a solid support.

A. The *p*-Nitrophenyl Ester Method

a. Synthetic Aspects

The *p*-nitrophenyl ester method was first applied for cyclization purposes by Schwyzer and Sieber in their brilliant synthesis of gramicidin S in 1957.[16] Since then, most analogs of gramicidin S have been synthesized by this cyclization method. We also used the *p*-nitrophenyl ester method in our initial synthesis of analogs of gramicidin S, with some modification as regards protecting groups. We used the *p*-methoxybenzyloxycarbonyl or *t*-butyloxycarbonyl group instead of the undesirably sensitive trityl group for α-amino protection and the benzyloxycarbonyl group instead of the *p*-toluenesulfonyl group for δ-amino protection of ornithine residues in the intermediate (**3.28**) in the synthesis of [Gly$^{5,5'}$]-gramicidin S.[77,78] The use of the benzyloxycarbonyl

Z(OMe)–(L-Val–L-Orn(Z)–L-Leu–D-Phe–Gly)$_2$–ONp

(3.28)

group for δ-amino protection provided some advantages over that of the *p*-toluenesulfonyl group in the final deprotection step, i.e., elimination of possible cleavage of acylproline peptide bonds[79–81] and of incorporation of inorganic salts, which are induced by treatment of the *p*-toluenesulfonyl group with sodium in liquid ammonia.

Di-*p*-nitrophenyl sulfite (**3.29**) is a useful reagent for preparing protected

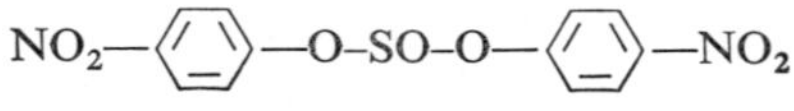

(3.29)

peptide *p*-nitrophenyl esters, e.g., the preparation of the acyldecapeptide *p*-nitrophenyl ester (**3.9**) with the reagent (**3.29**) shown in Fig. 3.1. However, the preparation of the reagent (**3.29**) is troublesome. Simplified procedures have been reported for this esterification as follows. Kawai and Nagai[82] treated a *t*-butyloxycarbonyl peptide acid with *p*-nitrophenyl trifluoroacetate to obtain the corresponding *p*-nitrophenyl ester. The *p*-nitrophenyl esters can also be prepared by the reaction of a protected peptide acid with dicyclohexylcarbodiimide and *p*-nitrophenol.[83–7)]

There is some risk of racemization of the *C*-terminal amino acid of a protected linear peptide acid during the preparation of its *p*-nitrophenyl ester. Thus, no optically active amino acids except for glycine and L- or D-proline, which are less prone to racemization, can be selected as the *C*-terminal amino acid of the protected linear peptide acid. To minimize the possible racemization, L-proline has been chosen in most cases as the *C*-terminus of a linear peptide *p*-nitrophenyl ester to be subjected to cyclization, and in some cases, glycine,[67,77,78,83,88–90] sarcosine,[91] β-alanine,[92] δ-aminovaleric acid,[93,94] and L-2-hydroxy-3-methylbutanoic acid[85] have been selected.

b. Cyclization of Linear Pentapeptide Esters

Around 1960, only five or so examples had been reported of the synthesis of cyclic pentapeptides.[95–9] However, it is not clear whether the cyclic pentapeptides synthesized were true monomers (cyclic pentapeptides) or dimers (cyclic decapeptides), because the techniques used for molecular weight determination of the cyclic peptides were not sufficiently precise at that time. In fact, several products obtained by cyclization of linear tripeptides and erroneously assigned as monomers (cyclic tripeptides)[100–2] were later proven to be dimerized cyclic hexapeptides.[103–5]

In 1966, Aoyagi and Izumiya[91] isolated a compound (**3.31**) less soluble in organic solvents in an attempt to obtain a dimer (**3.32**) (a cyclic decapeptide) by cyclodimerization of the pentapeptide *p*-nitrophenyl ester (**3.30**) (Fig. 3.10). The less soluble compound (**3.31**) was shown to be a cyclic monomer (a cyclic pentapeptide) as judged from various physicochemical properties.

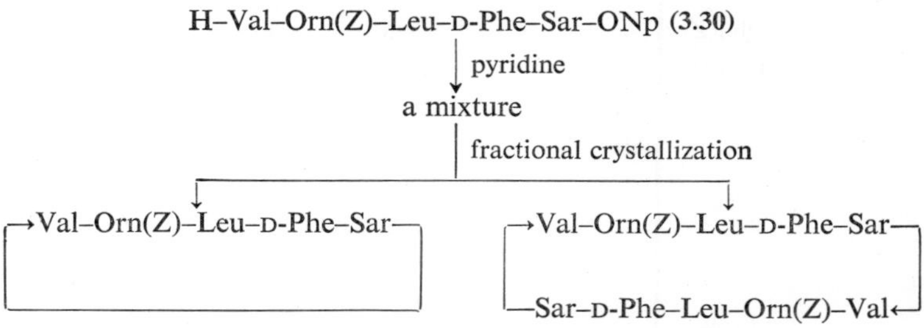

(**3.31**) 85 parts

(**3.32**) 15 parts

Fig. 3.10 Formation of monomer and dimer from a linear pentapeptide *p*-nitrophenyl ester.

Interestingly, the monomer (**3.31**) is less soluble than the dimer (**3.32**) in organic solvent, whereas the cyclic pentapeptide Z-semigramicidin S is more soluble than the cyclic decapeptide diZ-gramicidin S. The monomer (**3.31**) may be the first cyclic peptide definitely proved to be a cyclic pentapeptide by various physicochemical criteria.

This finding prompted us to examine the cyclization products obtained by treatment of linear pentapeptide *p*-nitrophenyl esters with pyridine. We came to recognize that there are three possibilities as regards the cyclization products. (1) Formation of a mixture of cyclic monomer and dimer, (2) monomer formation only, and (3) dimer formation only (Table 3.4). Table 3.4 shows that the ratio of formation of the cyclic monomer and the cyclic dimer depends on the sequence of each pentapeptide. In other words, the primary structure of the reacting pentapeptide active esters influences the mode of cyclization in pyridine and consequently the structure of the resulting cyclic

TABLE 3.4 Ratios of Protected Monomer and Dimer after Cyclization of Linear Pentapeptide *p*-Nitrophenyl Esters†1 (Source: ref. 2. Reproduced by kind permission of Kyoritsu Shuppan Co.)

No.	*p*-Nitrophenyl ester of pentapeptide†2					Ratio of compounds in product†3		Ref.
						Z-cyclic monomer	DiZ-cyclic dimer	
	1	2	3	4	5			
1	Val	Orn(Z)	Leu	D-Phe	Pro	32	68	17,18
2	Gly					100	0	106
3	Gly				Gly	100	0	88
4	Ala					91	9	106
5	Leu					78	22	107
6		Dbu(Z)				30	70	108
7		Lys(Z)				29	71	108
8			Gly			59	41	109
9			Ala			43	57	109
10				Gly		0	100	110
11				D-Ala		23	77	111
12				D-Val		17	83	112
13				D-Leu		15	85	112
14					Gly	79	21	67,77,78
15					Sar	85	15	91
16					βAla	89	11	92
17	Orn(Z)	Leu	D-Phe	Gly	Gly	100	0	88
18	D-Phe	Leu	Orn(Z)	Val	Pro	53	47	113,114

†1 The concentration of linear pentapeptide *p*-nitrophenyl ester in pyridine was approximately 3×10^{-3} M.

†2 Only the varied residues are shown except in the case of the first compound. All optically active amino acids are of the L-configuration unless otherwise indicated. Dbu, α,γ-diaminobutyric acid.

†3 The ratios shown are weight by weight.

peptides. In these cyclization experiments, a peptide *p*-nitrophenyl ester is reacted with pyridine at high dilution; the solution of the *p*-nitrophenyl ester is added drop by drop to a large amount of pyridine so as to reach a particular concentration of the ester (e.g., 3×10^{-3} M). Thus, on macroscopic considerations, the concentration of peptide *p*-nitrophenyl ester actually existing is presumed to be considerably lower than the selected final concentration (e.g., 3×10^{-3} M). However, on microscopic considerations, a higher concentration of the *p*-nitrophenyl ester than the final concentration is expected around the site of reaction of the active ester just after the addition of the active ester to pyridine. Table 3.4 shows the frequent occurrence of case 1 (formation of a mixture of cyclic monomer and dimer). In our experiments, only one example (No. 10 in Table 3.4) corresponds to the case of "dimer formation only" although Schwyzer *et al.*[64,65,115] suggested such preferential dimerization using pentapeptide *p*-nitrophenyl esters.

Let us consider the influence of each amino acid residue in the pentapeptide *p*-nitrophenyl ester (No. 1 in Table 3.4) on the cyclization reaction and products. Table 3.4 shows that the ratio of cyclic dimer decreases in the order L-valine, L-leucine and L-alanine substitution at position 1 of the pentapeptide (No. 1 in Table 3.4), and finally glycine substitution results in cyclic monomer exclusively. Thus the steric nature of the side chain of the *N*-terminal amino acid in the pentapeptide (No. 1 in Table 3.4) greatly influences the cyclization reaction. However, bulkiness or hydrophobicity of the amino acid side chain at position 3 has less effect. On the other hand, glycine substitution at position 4 produces only a dimer (a cyclic decapeptide) (No. 10 in Table 3.4). The presence of an L-proline residue at position 5 also has an important role in the dimerization reaction. In other words, the presence of amino acids with smaller side chains, such as glycine or L-alanine, at the *N*- or *C*-terminus of the pentapeptide *p*-nitrophenyl ester favors rapid peptide bond formation and consequent ring closure to monomers.

Schwyzer and Sieber reported that the cyclization of the pentapeptide *p*-nitrophenyl ester (**3.33**) afforded diMz-[$\text{Gly}^{1,1'}$, L-$\text{Lys}^{2,2'}$]-gramicidin S (**3.34**) in high yield.[115] No data on the molecular weight determination of

H–Gly–L-Lys(Mz)–L-Leu–D-Phe–L-Pro–ONp

Mz, *p*-methoxyphenylazobenzyloxycarbonyl

(**3.33**)

the diMz-compound which is designated as **3.34** were presented in their paper. We presume the cyclized product (**3.34**) obtained by Schwyzer and Sieber to be a monomer (a cyclic pentapeptide) by comparison with the result for No. 2 in Table 3.4 (monomer formation only). However, the presence of a bulky Mz group might hinder monomeric cyclization.

Let us consider the mode of the cyclization reaction of pentapeptide *p*-nitrophenyl esters (Table 3.4) at the molecular level. For this purpose we will adopt two simple conformational models for pentapeptide *p*-nitrophenyl esters in pyridine solution just before ring closure: one is of the type **3.35** and the other is of the type **3.36**, involving two molecules. The numbers in **3.35** and **3.36** correspond to those of the amino acid residues in gramicidin S. Table 3.4 shows that two conformational states such as **3.35** and **3.36** are

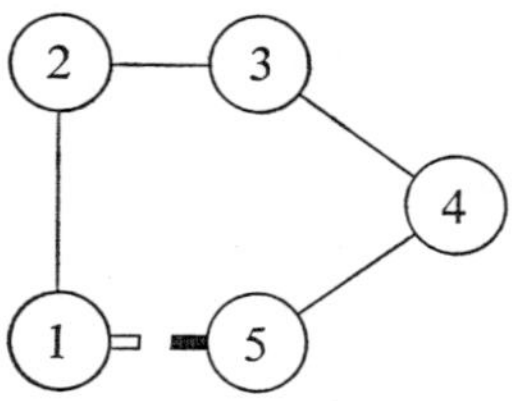

(3.35)

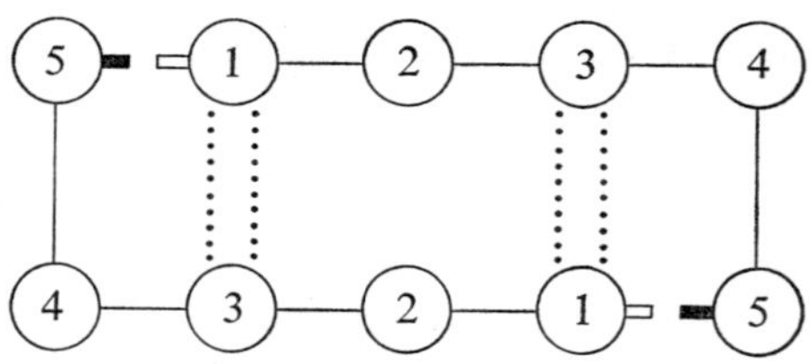

(3.36)

▭, amino group; ▬, active ester; . . . ., hydrogen bond
(Source: ref. 2. Reproduced by kind permission of Kyoritsu Shuppan Co.)

possible for pentapeptide esters having a D-amino acid at position 4 (Nos. 1, 11, etc. in Table 3.4). Monomer formation from such pentapeptide esters having a D-amino acid at position 4 reflects strong bending of the peptide chain around the D-amino acid portion, as shown in **3.35**. Glycine substitution at position 4 of the pentapeptide ester (No. 1 in Table 3.4) makes the peptide chain of the resulting molecule (No. 10 in Table 3.4) more flexible around that glycine portion, and intermediates of types **3.35** and **3.36** are both possible. However, the bulky side chain of L-valine at position 1 of the pentapeptide ester (No. 10 in Table 3.4) interferes with intramolecular peptide bond formation when the pentapeptide ester is in the form of **3.35**. Consequently, cyclodimerization of the pentapeptide ester (No. 10 in Table 3.4) occurs preferentially *via* **3.36**.

At present it is still difficult to predict the mode of cyclization of pentapeptide *p*-nitrophenyl esters related to gramicidin S because a great many combinations are possible for their amino acid sequences besides those shown in Table 3.4.

c. Cyclization of Linear Decapeptide Esters

Most of the gramicidin S analogs shown in Table 3.4 have been prepared by monomeric cyclization of linear decapeptide *p*-nitrophenyl esters as well as by cyclodimerization of pentapeptide *p*-nitrophenyl esters. In the cyclization system with decapeptide *p*-nitrophenyl esters, pure monomers (cyclic decapeptides) are usually obtained by recrystallization of the crude cyclized products. It is not clear whether or not cyclic polymers such as dimer (a cyclic eicosapeptide), trimer, etc. are formed, because the mother liquors from the recrystallizations have not been examined in detail. However, polymer formation seems to be very slight, if it occurs at all, in this cyclization system based on the results of chromatography of the recrystallization mother liquor. Let us consider the possible reason for the preferential cyclomonomerization of decapeptide *p*-nitrophenyl esters at the molecular level again. Decapeptide *p*-nitrophenyl esters have a tendency to form a structure similar to that of gramicidin S as shown in the model (**3.37**). Proximity of the α-amino group and *C*-terminal *p*-nitrophenyl ester in the model (**3.37**) results in the preferential formation of a cyclic monomer.

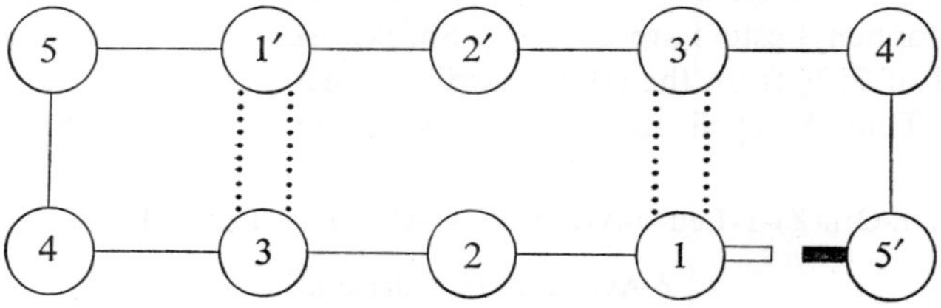

(3.37)

(Source: ref. 2. Reproduced by kind permission of Kyoritsu Shuppan Co.)

Kawai and Nagai[82,116] reported the formation of cyclic polymers from a linear decapeptide *p*-nitrophenyl ester containing α-aminoisobutyric acid as follows. Treatment of the compound (**3.38**) with pyridine afforded a mixture of three cyclic products with different molecular sizes. One of them was ascertained as a cyclic monomer (a cyclic decapeptide), and the other products

H–(L-Val–L-Orn(Z)–L-Leu–Aib–L-Pro)$_2$–ONp

Aib, α-aminoisobutyric acid

(3.38)

H–(L-Val–L-Orn(Z)–L-Leu–D-Phe–Aib)$_2$–ONp or –ONSu

(3.39) or **(3.40)**

H–(L-Val–L-Orn(Z)–L-Leu–D-Phe–Aib)$_2$–OCH$_2$CN

(3.41)

were presumed to be a dimer (a cyclic eicosapeptide) and a trimer (a cyclic triacontapeptide). On the other hand Takiguchi *et al.*[117] reported an unsuccessful attempt to cyclize the decapeptide active ester (**3.39** or **3.40**) containing α-aminoisobutyric acid under usual cyclization conditions, but succeeded in synthesizing the desired [α-aminoisobutyric acid$^{5,5'}$]-gramicidin S by cyclization of a different type of active ester (**3.41**). An unusually sterically hindered side chain of α-aminoisobutyric acid in these peptides, **3.38**, **3.39** and **3.40** leads to unusual features in the cyclization products or cyclization reaction itself. Thus the presence of α-aminoisobutyric acids at positions 4 and 4′ in **3.38** may force the peptide chain of **3.38** to straighten around the regions of these unusual amino acids, leading to the production of cyclic dimer and trimer.

The yield in cyclization reactions of decapeptide *p*-nitrophenyl esters will be touched on briefly. Generally cyclomonomerization yields from decapeptide *p*-nitrophenyl esters are 20–30%. Sōfuku isolated a pure monomer in a high yield of 71% from the corresponding linear peptide *p*-nitrophenyl ester (**3.42**).[118] Thus the cyclization yield depends on the amino acid sequence of

H–L-Val–L-Orn(Z)–L-Leu–δ-Ava–L-Val–L-Orn(Z)–L-Leu–D-Phe–L-Pro–ONp

δ-Ava, δ-aminovaleric acid

(3.42)

the starting linear peptide *p*-nitrophenyl esters and also on the experimental conditions used.

B. The *N*-Hydroxysuccinimide Ester Method

Use of the *N*-hydroxysuccinimide ester for cyclization has some advantages over that of the traditional *p*-nitrophenyl ester described above. For example, cyclization with the *N*-hydroxysuccinimide ester proceeds under mild conditions in a fairly high yield (50–70%). The *N*-hydroxysuccinimide ester method was used for an initial fundamental study[68] and subsequently for the syntheses of macro-ring analogs of gramicidin S,[119,120] gramicidin S analogs containing *N*-methyl-L-leucine or *N*-methyl-L-valine,[121–3] [L-Val$^{3,3'}$]-gramicidin S,[124] [Gly$^{5,5'}$]-gramicidin S,[125] [L-Ala$^{1,1',3,3'}$]-gramicidin S[126] and [L-Phe$^{1,1',3,3'}$]-gramicidin S.[127] This method was also used in the synthesis of cyclic depsipeptides, AM-toxin I and its analogs,[128] which will be described in more detail in Chapter 8.

Let us consider the synthesis of macro-ring analogs of gramicidin S at the molecular level. Cyclization of the linear peptide *N*-hydroxysuccinimide ester (**3.43**) ($n = 3$ or 4) in pyridine afforded triZ-sesquigramicidin S or tetraZ-digramicidin S, respectively, in high yield (about 50%) without the formation of cyclic polymers. The high yields of these cyclomonomerization

H–(L-Val–L-Orn(Z)–L-Leu–D-Phe–L-Pro)$_n$–ONSu

(3.43)

products from the linear peptide active esters (**3.43**) (n = 3 and 4) can be explained in terms of the following conformational states of the active esters (**3.43**) (n = 3 and 4) just before ring closure (**3.44** and **3.45**). This view is

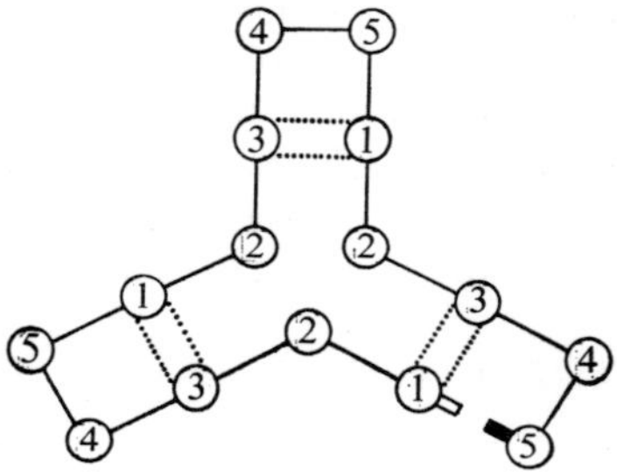

(3.44)

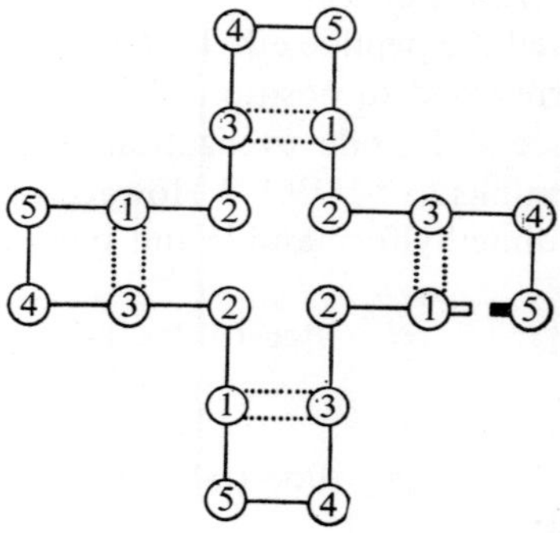

(3.45)

(Source: ref. 2. Reproduced by kind permission of Kyoritsu Shuppan Co.)

favored by the results of experiments on the optical rotatory dispersion of naked linear polypeptides (**3.46**) (n = 3 and 4) or cyclic sesqui- and digramicidin S (**3.47**) (n = 3 and 4).[119,120] The linear and macro-ring analogs, **3.46** and

H–(L-Val–L-Orn–L-Leu–D-Phe–L-Pro)$_n$–OH

(3.46)

cyclo(–L-Val–L-Orn–L-Leu–D-Phe–L-Pro–)$_n$

(3.47)

3.47, as well as gramicidin S have similarly shaped curves with a negative trough at 232 nm in ethanol.

C. The Azide Method

Recently increasing numbers of papers have appeared dealing with cyclization by the azide method. For example, the method was used in a fundamental study[68] and in syntheses of gramicidin S,[70,74,76] [Gly$^{5,5'}$]-gramicidin

S,[129] [L-Phe$^{5,5'}$]-gramicidin S,[130] [L-His^{2}]- and [L-His$^{2,2'}$]-gramicidin S,[131] gramicidin S analogs containing L-serine, glycine or sarcosine,[132] and gramicidin S analogs containing *S*-alkyl-L-cysteine.[133] Today the azide method is still one of the best racemization-free coupling procedures. Thus it is possible to select any optically active amino acid as the *C*-terminal amino acid of a linear peptide just before ring closure using this method, as is also the case with the recently developed *N*-hydroxysuccinimide ester method described in the preceding section.

In an attempt to synthesize [L-Leu$^{5,5'}$]-gramicidin S, Abe *et al.*[134] prepared the peptide ethyl esters (**3.48**) ($n = 1$ and 2; their *C*-terminal L-leucines correspond to positions 3 or 3′ in the gramicidin S sequence) expecting the successful azide cyclization of linear peptides with L-leucine as their *C*-terminus.[70,74,76,129,133] However, the esters (**3.48**) were sparingly soluble even in dimethylformamide and only partial hydrazinolysis of **3.48** occurred. By

Boc–(D-Phe–L-Leu–L-Val–L-Orn(Z)–L-Leu)$_n$–OEt

(**3.48**)

Boc–(L-Val–L-Orn(Z)–L-Leu–D-Phe–L-Leu)$_2$–OEt

(**3.49**)

trial and error Abe *et al.* synthesized the decapeptide ester (**3.49**) (its *C*-terminal L-leucine corresponds to position 5′ in the gramicidin S sequence) which was more soluble in organic solvent than **3.48** and could be successfully converted into the corresponding hydrazide to afford the desired dibenzyloxycarbonyl cyclic analog *via* the azide cyclization method.

D. The Phosphite Method

Rothe *et al.* cyclized linear pentapeptides by the phosphite method to obtain several gramicidin S analogs[63,135] as shown in Table 3.5, and other analogs such as all-L-gramicidin S[63,135] and [L-Lys$^{2,2'}$, Gly$^{4,4'}$]-gramicidin S,[135,136] although an elevated temperature (100° C) was employed. Table 3.5 indicates that pentapeptide active esters (mixed anhydride type) formed in diethyl phosphite by this phosphite method are mainly of the type **3.35**, leading to predominant formation of cyclic monomers compared with the *p*-nitrophenyl ester–pyridine system. As shown in Table 3.3, the ratio of the cyclic monomer and the cyclic dimer is almost the same after cyclization of linear pentapeptides regardless of the activation method (*p*-nitrophenyl ester, *N*-hydroxysuccinimide ester or azide).[68] Thus the nature of the solvent used seems to effect the ease of formation of the intermediate, **3.35** or **3.36**, rather than the type of activation.

TABLE 3.5 Ratios of Protected Monomer and Dimer after Cyclization of Linear Pentapeptides by Two Activation Methods (Source: ref. 2. Reproduced by kind permission of Kyoritsu Shuppan Co.)

Amino acid sequence of peptide portion of the active esters†1					Weight ratio of compounds in product (By *p*-nitrophenyl ester method†3)		(By phosphite method)	
					Monomer	Dimer	Monomer	Dimer
Val	Orn(R†2)	Leu	D-Phe	Pro	32	68	62	38
			Phe		—	—	40	60
			Gly		0	100	80	20

†1 Only the varied residues are shown except in the case of the first compound. All optically active amino acids are of the L-configuration unless otherwise indicated.
†2 R = Z, Boc or Pht.
†3 From Table 3.4.

E. The Dicyclohexylcarbodiimide–*N*-Hydroxysuccinimide Method

Halstrøm and Klostermeyer[137] synthesized [Gly$^{5,5'}$]-gramicidin S *via* excess dicyclohexylcarbodiimide-mediated cyclization using excess *N*-hydroxysuccinimide instead of *p*-nitrophenol as an additive in procedure similar to the synthesis of gramicidin S (see Fig. 3.7). This cyclization method may well come to be used widely because of its simplicity and convenience.

F. The Acid Chloride Method

The acid chloride method is rarely used in the synthesis of homodetic cyclic peptides because use of the highly reactive acid chloride leads to extensive racemization of the *C*-terminal amino acid of linear peptides just before ring closure. However, the use of this method is apparently acceptable in the synthesis of cyclic depsipeptides if an α-hydroxy acid is selected as the *C*-terminal component of a linear depsipeptide. Thus a free linear decadepsipeptide with L-2-hydroxy-3-methylbutanoic acid as its *C*-terminus was activated with the aid of thionyl chloride to afford a unique cyclodepsipeptide analog of gramicidin S, [L-2-hydroxy-3-methylbutanoic acid$^{1,1'}$, L-2-hydroxy-4-methylpentanoic acid$^{3,3'}$]-gramicidin S in a yield of 13%.[85]

G. Cyclization on a Solid Support

Peptide cyclizations are generally carried out at high dilution to prevent intermolecular and to promote intramolecular reactions. The use of a solid support as a carrier for the molecule to be cyclized seems an ideal system for cyclizations. In this system the dilution is almost infinitely high in principle, so that only intramolecular cyclizations are expected to occur, resulting in the formation of cyclic monomers. Thus the system may be effective for obtaining semigramicidin S (a cyclic pentapeptide) in quantity. In fact, Fridkin *et al.*[138)] prepared a linear peptide active ester linked with an insoluble support (**3.50**) from which monomeric cyclo(–L-Ala–L-Ala–L-Ala–L-Ala–), normally inaccessible, was obtained in a high yield. Recently a new solid support bearing *N*-hydroxysuccinimide[139)] or 1-hydroxybenzotriazole[140)] moieties has been developed. High reactivity of peptide active esters linked

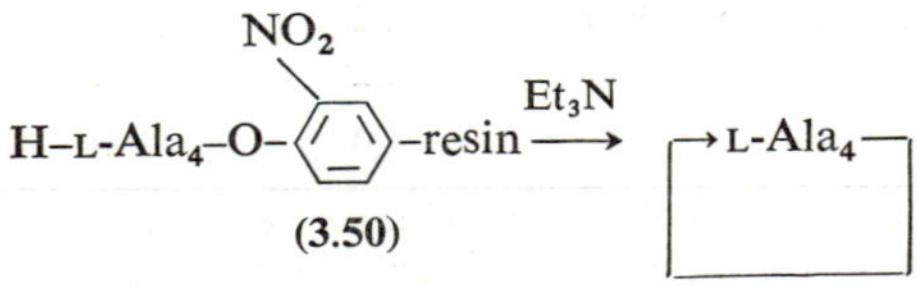

(**3.50**)

with these supports is anticipated.

Flanigan and Marshall[141)] combined the solid-phase and active ester methods as shown in Fig. 3.11. Prior to cyclization, the resin-bound peptide ester (**3.52**) was activated by oxidation to the corresponding active ester (**3.53**). A cyclic tetrapeptide was obtained in an overall yield of 40% from the starting resin-bound amino acid ester (**3.51**).

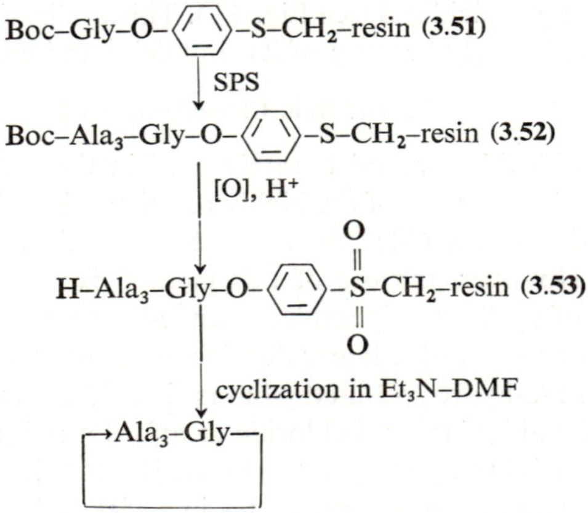

Fig. 3.11 Synthesis of a cyclic tetrapeptide on a solid support. SPS, solid-phase synthesis.

Attachment of a peptide to a solid support through the side chain of a trifunctional amino acid residue is another interesting approach to obtain cyclic peptides. For example, Sklyarov and Shashkova[125)] succeeded in the

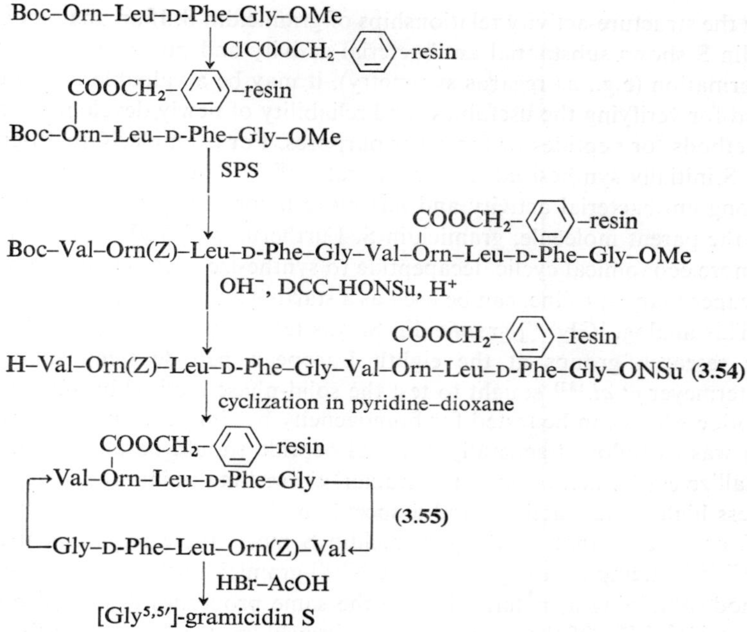

Fig. 3.12 Synthesis of [Gly$^{5,5'}$]-gramicidin S on a solid support. SPS, solid-phase synthesis.

synthesis of [Gly$^{5,5'}$]-gramicidin S by utilizing the δ-amino group of ornithine as an anchoring site. The synthetic scheme employed is shown in Fig. 3.12. Cyclization of the resin-bound decapeptide active ester (**3.54**) furnished the resin-bound cyclic peptide (**3.55**). Removal of the solid support from **3.55** with hydrogen bromide in acetic acid afforded [Gly$^{5,5'}$]-gramicidin S in 20% overall yield based on the amount of tetrapeptide initially attached to the resin.

3.2.2 SYNTHETIC ASPECTS OF ANALOGS OF GRAMICIDIN S

Several synthetic aspects of gramicidin S analogs will be considered in this section: gramicidin S analogs as model peptides, gramicidin S analogs containing *N*-methylamino acids, hybrid analogs of gramicidin S and derivatives of natural gramicidin S.

A. Analogs of Gramicidin S as Model Peptides

One of the main reasons for the synthesis of gramicidin S analogs is to study the structure-activity relationships of gramicidin S. If an analog of gramicidin S shows substantial antibacterial activity and possesses a favorable conformation (e.g., as regards symmetry), it may be a valuable model compound for verifying the usefulness and reliability of newly developed synthetic methods for peptides or for other purposes. For example, [Gly$^{5,5'}$]-gramicidin S initially synthesized in our laboratory[67,77,78] unexpectedly exhibited a strong antibacterial activity and has more favorable symmetry properties than the parent molecule, gramicidin S. Furthermore, [Gly$^{5,5'}$]-gramicidin S is a more economical cyclic decapeptide to synthesize because glycine, which is cheaper than L-proline, can be used as a starting amino acid.

This analog, [Gly$^{5,5'}$]-gramicidin S, was taken up as a model peptide by three research groups at the eighth European peptide symposium.[142-4] Klostermeyer *et al.*[142] sought to test the solid-phase method by synthesizing a peptide which can be tested for homogeneity by simple means. Their argument was as follows: generally a model peptide should be of medium size, crystallize easily, i.e., possess a characteristic melting point, and also should possess high optical activity and a specific biological activity. Klostermeyer *et al.* considered that [Gly$^{5,5'}$]-gramicidin S meets these requirements, and they[137,142] attempted to synthesize [Gly$^{5,5'}$]-gramicidin S by the solid-phase method, obtaining a material having the same properties described by Aoyagi *et al.*[67,77,78] Of the other groups, Kiryushkin *et al.*[83,143] used the same analog as a model peptide for testing a soluble polymer support, and Ovchinnikov *et al.*[90,144-7] used it for topochemical purposes. Sklyarov and Shashkova[125] and Halstrøm *et al.*[129] also utilized this analog as a model peptide to confirm the usefulness of their modified solid-phase methods.

B. Analogs of Gramicidin S Containing *N*-Methylamino Acids

We have mainly discussed cyclization so far because there are not many problems in developing a synthetic strategy for building up a linear peptide chain. In this section we will give examples where care is required in the synthesis of linear peptides.

The synthesis of analogs containing *N*-methylamino acids provides an interesting and important means to study the structure-activity relationships not only of gramicidin S but also of biologically active peptides. However, some problems are encountered in synthesizing a peptide containing an *N*-methylamino acid, e.g., the lack of crystallizability of the intermediate, the lower reactivity of the methylamino group, etc. In spite of these difficulties, interesting analogs such as [L-MeLeu^{3}]- and [L-MeLeu$^{3,3'}$]-gramicidin S have been synthesized.[121,122] The synthetic route for a pentapeptide intermediate (**3.56**) containing *N*-methyl-L-leucine is shown in Fig. 3.13. Later we noted the possibility of racemization of the *N*-methyl-L-leucine residue in this synthetic strategy (Fig. 3.13), because McDermott and Benoiton[148,149]

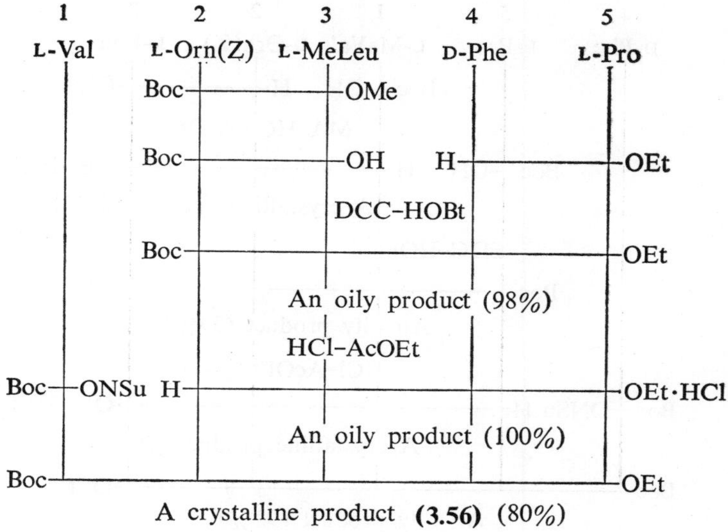

Fig. 3.13 Synthetic route for a pentapeptide containing L-MeLeu.

pointed out that a widely used and relatively racemization-free coupling reaction caused partial racemization of the *N*-methylamino acid residue of the tripeptide **(3.58)** that was prepared from the protected dipeptide **(3.57)** con-

$$\underset{(\mathbf{3.57})}{\text{Z–L-Ala–L-MeLeu–OH}} + \text{H–Gly–OBzl} \xrightarrow{\text{DCC–HONSu}} \underset{(\mathbf{3.58})}{\text{Z–L-Ala–MeLeu–Gly–OBzl}}$$

(Extent of racemization of the MeLeu residue in **3.58**, 3–11%).

taining *N*-methyl-L-leucine. We also examined the coupling reaction of Z–Gly–L-MeAla–OH and H–L-Leu–OBzl by the dicyclohexylcarbodiimide–1-hydroxybenzotriazole (or *N*-hydroxysuccinimide) method and confirmed that partial racemization of the *N*-methylalanine residue of the resulting tripeptide had occurred.

In the synthesis of the pentapeptide **(3.56)** by Sugano *et al.*[121,122] it was fortunate that the peptide **(3.56)** had good crystallinity, as shown in Fig. 3.13. Thus a small quantity of diastereomer that might be induced by partial racemization of the *N*-methylleucine residue could be eliminated by recrystallization of the crude pentapeptide **(3.56)**. In recent syntheses of [L-MeVal^{1}]-gramicidin S and [L-MeVal$^{1,1'}$]-gramicidin S, Abe *et al.*[123] employed a stepwise elongation method (Fig. 3.14) that was free from the risk of racemization.

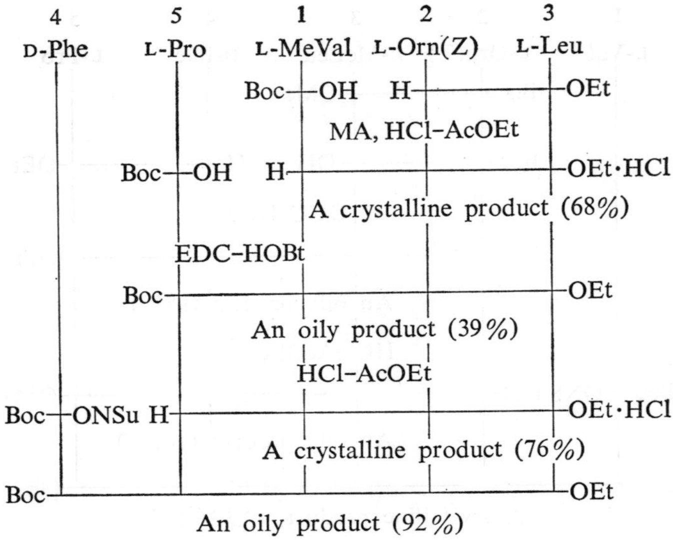

Fig. 3.14 Synthetic route for a pentapeptide containing L-MeVal.

C. Hybrid Analogs of Gramicidin S

Gramicidin S consists of two repeated pentapeptide sequences. Thus an analog of gramicidin S in which one or more amino acids in one of these two pentapeptide sequences are replaced by other amino acids might be regarded as a hybrid analog of gramicidin S. [Gly^5]-gramicidin S was the first synthesized hybrid analog of gramicidin S.[89] The synthesis of these hybrid analogs of gramicidin S has to be carried out *via* monomeric cyclization of the corresponding activated decapeptide. A cyclic nonapeptide (**3.59**) reported by Sōfuku[118] can also be regarded as a hybrid analog of gramicidin S.

1 2 3 4–5 1′ 2′ 3′ 4′ 5′
cyclo(–L-Val–L-Orn–L-Leu–δ-Ava–L-Val–L-Orn–L-Leu–D-Phe–L-Pro–)

δ-Ava, δ-aminovaleric acid

(3.59)

D. Derivatives of Natural Gramicidin S

Natural gramicidin S is now commercially available. Simple chemical modification of an appropriate part of the molecule of gramicidin S can afford a unique analog of gramicidin S whose preparation is not easy *via* stepwise chemical synthesis. The chemically reactive sites in the molecule of gramicidin S are the δ-amino groups of ornithine residues, benzene rings of the side chains of phenylalanine residues and eight –CONH– groups of the peptide backbone of the molecule. The disadvantage of this modification approach is the limited number of modifiable sites in the molecule of gramicidin S, whereas stepwise chemical synthesis can offer a wide variety of analogs of gramicidin S. As an initial study along this line of chemical modification, a Russian group reported the synthesis and antibacterial activity of several derivatives of natural gramicidin S, such as [L-Orn(Gly)$^{2,2'}$]-gramicidin S, obtained by coupling the two δ-amino groups of gramicidin S with Z-glycine and subsequent hydrogenation.[150] Recently Nishino *et al.*[151] prepared [L-Arg$^{2,2'}$]-gramicidin S by treating natural gramicidin S with *S*-methylisothiourea. In their synthesis of [L-Arg$^{2,2'}$]-gramicidin S it was necessary to use a large excess of the reagent, *S*-methylisothiourea, and to use a long reaction time to ensure the completion of the guanidination.

Modification of one of the two δ-amino groups of gramicidin S gives an interesting natural hybrid analog suited for studies on the antibacterial and physicochemical properties of gramicidin S. Minari *et al.* in our laboratory succeeded in the preparation of a large amount of monoZ-gramicidin S, from which several monoacyl-gramicidin S compounds **(3.60)** (e.g., monoacetyl-gramicidin S) were obtained, as shown in Fig. 3.15. MonoBoc-gramicidin

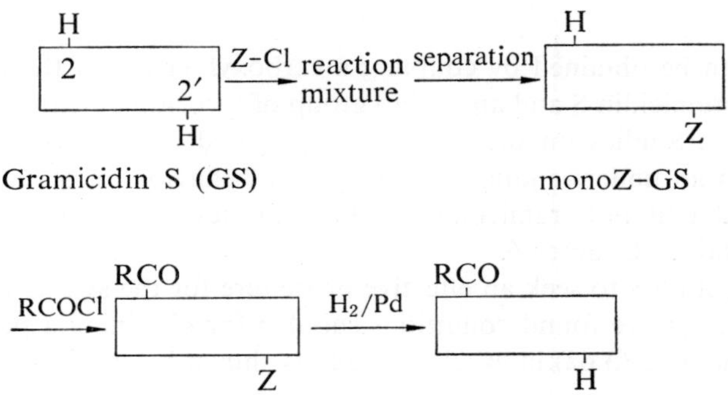

(3.60)

Fig. 3.15 Preparation of monoacyl-gramicidin S.

S,[152] obtained by selective deprotection of diBoc-gramicidin S, is also an important starting material for derivatives of natural gramicidin S. Waki *et al.* in our laboratory intend to prepare an "aggregated analog" of gramicidin S **(3.61)** from this monoBoc-gramicidin S by successive coupling with *N*-succinyl-monoZ-gramicidin S and subsequent deprotection. Another interesting aggregated analog which contains gramicidin S and tyrocidine in one

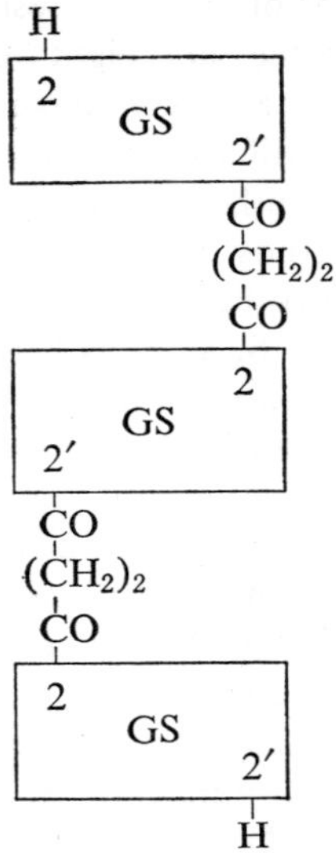

GS, gramicidin S

(3.61)

molecule can be obtained by coupling a carboxyl group of the *N*-succinyl-monoacyl-gramicidin S and an amino group of tyrocidine with dicyclohexyl-carbodiimide. Studies on these unique aggregated analogs of gramicidin S may lead to an understanding of aggregation behavior or of the mode of antibacterial action of gramicidin S and tyrocidines, which will be disscussed in more detail in Chapter 6.

During studies to seek an effective procedure for the synthesis of *N*-methylamino acids, we found conditions suitable for obtaining a protected *N*-methylamino acid (**3.62**) in excellent yield as shown below.[153] Tanaka *et al.*

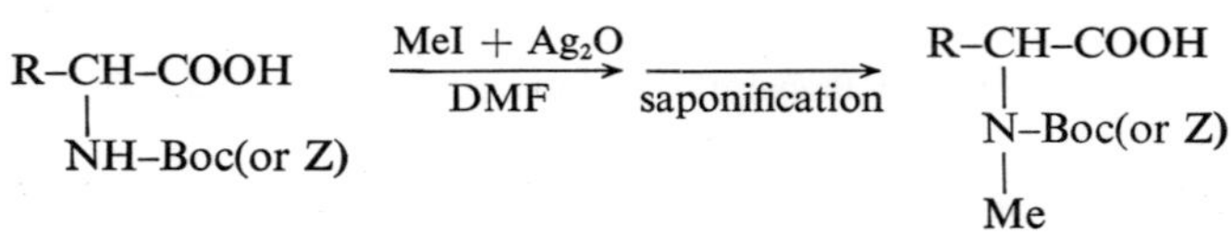

(3.62)

in our laboratory utilized these *N*-methylation conditions for the *N*-methylation of diZ-gramicidin S and followed the time course of *N*-methylation (–CO–

NH– ⟶ –CO–NMe–) of the peptide backbone of diZ-gramicidin S. Using this technique, we can determine the influence of *N*-methylation of gramicidin S derivatives on their conformation and antibacterial activity. We assumed that four amide protons (–NH–) of peptide bonds participating in the four hydrogen bondings of diZ-gramicidin S might be resistant to *N*-methylation on the basis of the conformational model of gramicidin S (see Fig. 3.3). In fact, the content of *N*-methylvaline and *N*-methylleucine were smaller than that of *N*-methylphenylalanine in acid hydrolysates of the *N*-methylated diZ-gramicidin S. The results indicate that the four amide protons of valine and leucine residues of diZ-gramicidin S may participate in hydrogen bonding in the molecule.

Recently Kawai and Nagai[116,154] reported a similar approach using diphthaloyl or di-dinitrophenyl-gramicidin S as a starting material. Their starting compounds are superior to our diZ-gramicidin S, in which δ-amino hydrogens may still be methylated by the *N*-methylation procedure. They further applied the *N*-methylation procedure to several analogs of gramicidin S, such as di-dinitrophenyl-[α-aminoisobutyric acid$^{4,4'}$]-, [L-Ala$^{4,4'}$]- and [D-Ala$^{4,4'}$]-gramicidin S, and determined the positions and extent of hydrogen bonding by following the time courses of quantitative amino acid analyses of the acid hydrolysates of the methylated samples.[82,116] Thus, this *N*-methylation method provides a simple and effective chemical approach which can be used to estimate the positions and extent of hydrogen bonding in gramicidin S derivatives, various analogs of gramicidin S, and also biologically active peptides.

3.3 Synthesis of Tyrocidines and Their Analogs

Tyrocidines A–E are natural homologs of tyrocidine. They can also be regarded as natural hybrid analogs of gramicidin S, as can be seen from their structures (see Fig. 2.1). Thus the synthesis of these tyrocidines has to be carried out *via* the cyclization of corresponding activated linear decapeptides. The preparation of tyrocidine homologs and their synthetic analogs has been a major interest of our laboratory. In 1966, Ohno *et al.*[155,156] succeeded in synthesizing tyrocidine A for the first time by allowing the corresponding decapeptide *p*-nitrophenyl ester **(3.63)** to cyclize. Since then Kuromizu *et al.*

H–L-Phe–D-Phe–L-Asn–L-Gln–L-Tyr–L-Val–L-Orn(Z)–L-Leu–D-Phe–L-Pro–ONp

(3.63)

and Mitsuyasu *et al.* have prepared tyrocidines B,[157,158] C,[159,160] and E[161,162] by the same type of cyclization reaction. Recently Okamoto *et al.*[163] synthesized tyrocidine A and various analogs by the solid-phase method, employing a synthetic route similar to that shown in Fig. 3.8, in which L-leucyl-resin was used as a starting material. They also synthesized tyrocidine A and its analogs ([Gly^6]- and [Gly^7]-tyrocidine A) by the solution method.[164] The same linear decapeptide sequence as that used in the solid-phase synthesis was employed, in which L-leucine at position 3 of tyrocidine A occupies the *C*-terminus. Okamoto *et al.*[165] also found suitable conditions for separating a mixture of natural tyrocidine homologs by a new technique of countercurrent chromatography, namely droplet countercurrent chromatography. They applied this new technique for purification and confirmation of the purity of synthetic tyrocidine A and its analogs obtained by the solution method.[164]

Our synthetic design for tyrocidine A analogs was focused on replacement of the amino acid residues at positions 6 to 10 in the tyrocidine A molecule by glycine. The simplest such analog is of the type **3.64**. A similar compound is also the simplest type of hybrid analog of gramicidin S.

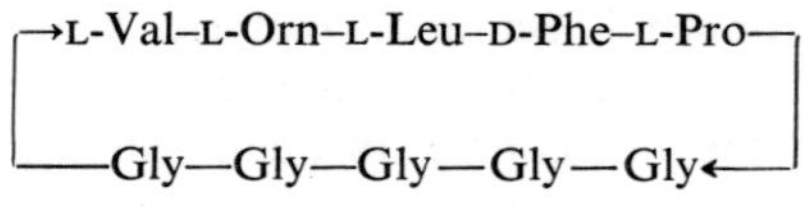

(3.64)

Easily obtainable modified analogs of natural tyrocidines can be utilized for physicochemical studies of tyrocidines. For example, hydrogenated tyrocidines A, B and C were used for CD studies of tyrocidines.[166] Ruttenberg *et al.*[167] prepared several interesting modified analogs of tyrocidines for studies on association behavior: open-chain tyrocidine B, *N*-succinyl tyrocidine B and a condensation product (**3.65**) of tyrocidine B and *N*-succinyl-

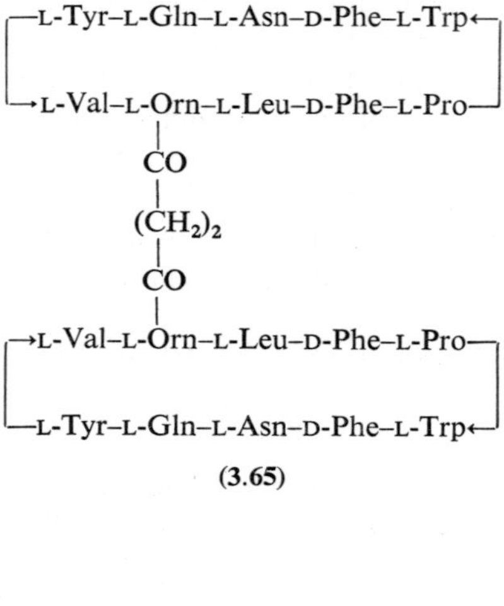

(3.65)

tyrocidine B obtained using dicyclohexylcarbodiimide. The condensation product (**3.65**) is regarded as an aggregated analog of tyrocidine B.

CHAPTER **4**

Relationship between the Primary Structure and Activity of Gramicidin S and Tyrocidines

For an understanding of the structure-activity relationships of biologically active peptides, the use of a variety of synthetic variants of the natural peptides is the most powerful weapon. In fact such a synthetic approach to structure-activity correlations has produced significant progress in the cases of many peptide hormones (e.g., oxytocin, adrenocorticotropin, luteinizing hormone-releasing hormone, somatostatin, and so on). The same approach has been applied to peptide antibiotics. However, most peptide antibiotics have different structural features compared with peptide hormones: a cyclic structure, an unusual amino acid such as a D-amino acid, hydroxy acid or other component as a constituent. These special characteristics of peptide antibiotics lead to difficulties in the syntheses of parent antibiotics and their analogs, and have limited synthetic work on them. Thus, elucidation of the relationship between the structure and activity of gramicidin S by the synthetic approach is very important as a model study for general peptide antibiotics not only from a fundamental but also from a practical point of view.

Gramicidin S is one of the peptide antibiotics that have been most extensively studied by the synthetic approach. In designing practically useful derivatives of naturally occurring peptides, it is desirable to develop more potent or long-acting biological activity and a broader biological spectrum compared with those of the natural paren peptides. Although a wide variety of analogs of gramicidin S have been synthesized so far, little attention has been paid to developing analogs with these properties. One reason is that considerable time and expense are necessary for the synthesis of analogs of gramicidin S on a factory scale, whereas gramicidin S can be produced easily from bacteria for medical use. However, such fundamental studies as chemical modification of natural gramicidin S and synthesis of polyamino acids with antibacterial activity resembling that of gramicidin S may be worthwhile as regards practical considerations.

In this chapter our attention will be focused on variations of the primary (planar) structure of gramicidin S and of the resulting activity. For a better understanding of the function of gramicidin S, it may be necessary to determine the mode of interaction between gramicidin S and sensitive bacteria and to analyze the properties of receptor sites of the bacteria. These points will be covered in more detail in Chapter 6. In some cases we will discuss the correlation of the structure and activity of gramicidin S based on an assumed higher (secondary and tertiary) structure of gramicidin S and its analogs and with assumptions regarding possible receptor sites in the sensitive cell membrane. Our working hypothesis (Fig. 4.1) concerning the interaction between gramicidin S or its active analogs and gramicidin S-sensitive bacteria is presented as an aid to discussion. The numbers in Fig. 4.1 indicate the amino acid residues in gramicidin S, as mentioned in Section 3.2.

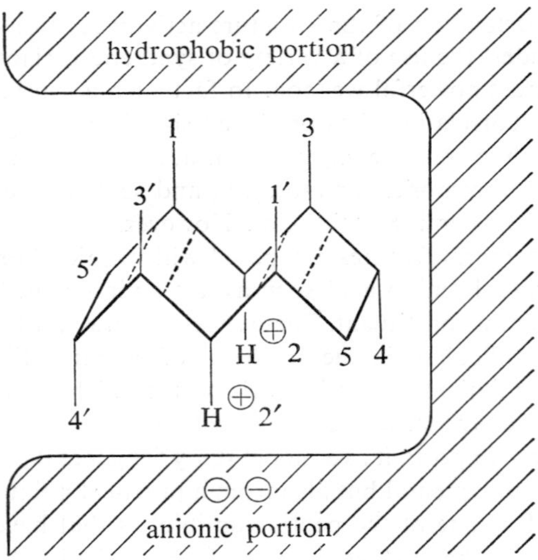

Fig. 4.1 Interaction between gramicidin S or its active analogs and gramicidin S-sensitive bacteria. (Source: ref. 2. Reproduced by kind permission of Kyoritsu Shuppan Co.)

Although the antibacterial activity of test samples is influenced by the test method and the conditions used, we use here the definition that a peptide has no activity when it does not inhibit the growth of microorganisms even at the level of 100μg/m*l*. Gramicidin S itself inhibits the growth of Gram-positive bacteria (*Bacillus subtilis, Staphylococcus aureus,* etc.) at levels of 1–10 μg/m*l*.

4.1 Linear Peptides Related to Gramicidin S

Linear peptides ranging in size from oligopeptides to polypeptides related to gramicidin S will be described in this section.

4.1.1 LINEAR OLIGOPEPTIDES

As soon as the primary structure of gramicidin S had been elucidated, several groups began to study the antibacterial properties of partial peptide fragments of gramicidin S. The presence of an amino acid of D-configuration, D-phenylalanine, and another amino acid, L-ornithine, not encountered in proteins appeared to be characteristic of the antibiotic. Secondly, the fact that the same pentapeptide sequence, –Val–Orn–Leu–D-Phe–Pro–, was present in tyrocidines suggested that this amino acid sequence itself was related to the specific antibacterial activity of gramicidin S.

As early as 1950, Harris and Work[168,169] synthesized several linear oligopeptides ranging in size from dipeptides to pentapeptides. One of them is shown below. However, none of them proved to be active. In 1954,

H–L-Val–L-Orn–L-Leu–D-Phe–L-Pro–OH

Erlanger and Goode reported an important result in their initial studies on gramicidin S analogs. Namely, the synthetic linear decapeptide analog (**4.1**)[170] of gramicidin S exhibited about one-tenth of the specific activity of gramicidin S (Table 4.1).[171] Later they also reported that newly synthesized linear decapeptides (**4.2** and **4.3**) showed weak activity (Table 4.1).[172] As shown in Table 4.1 these linear decapeptides exhibit synergistic activity with gramicidin S. The mode of antibacterial action of these linear decapeptides may thus

TABLE 4.1 Antimicrobial Activities of Linear Decapeptides and Their Mixtures with Gramicidin S (Source: ref. 2. Reproduced by kind permission of Kyoritsu Shuppan Co.)

Peptide†1	Minimum inhibitory concentration (μg/ml) *E. coli*
Gramicidin S (GS)	1.85
H–(Val–Orn–Leu–D-Phe–Pro)$_2$–OH (**4.1**)	15
H–(D-Tyr)$_2$–OH(**4.2**)†2	107
H–(Lys)$_2$–OH(**4.3**)†2	66.7
4.1 + GS	0.5 + 3.8
4.3 + GS	0.5 + 12.8

†1 All optically active amino acids are of the L-configuration unless otherwise indicated.
†2 Only the varied residues are shown except in the case of compound (**4.1**).

differ from that of gramicidin S. In any event, the results imply that the cyclic structure is important for strong activity but not indispensable for the appearance of antibacterial activity of gramicidin S. It should be pointed out here that the compound (**4.1**) is one of five possible linear decapeptides with the sequence of gramicicin S.

In 1968, Pollard *et al.*[173)] observed the formation of a formyl decapeptide ethanolamide (**4.4**) (Table 4.2) in a cell-free biosynthetic system for gramicidin S. However, they did not describe any of the physical or biological properties of this possible immediate precursor (**4.4**) of gramicidin S, probably because of the small amount isolated. Therefore Makisumi *et al.* synthesized the corresponding peptide (**4.4**)[174,175)] and related linear pentapeptide[176)] and decapeptides (**4.1** and **4.5**)[177)] and tested their antibacterial activity. Makisumi *et al.* reported that none of the pentapeptides showed activity, and that the amino acid sequence of the linear decapeptides had some influence on their activity (Table 4.2). The results in Table 4.2 were unexpected to us. At first we presumed that the order of increasing specific activity of the linear decapeptides would be **4.1**, **4.5** and the natural precursor (**4.4**), but in fact opposite results were obtained. The reason for this may be as follows. Gramicidin S, which has a β-sheet structure (see Fig. 3.3 in Chapter 3), can bind to the receptor sites of sensitive bacteria and inhibit the growth of the bacteria (also see Fig. 4.1). The related compound (**4.1** or **4.5**) also forms a similar spatial structure of the type **3.37** described in Section 3.2.1 and can fit to the receptor sites of bacteria, thus exhibiting activity. However, compound (**4.4**) may not adopt such a spatial conformation and may bind to the receptor sites of bacteria only weakly.

TABLE 4.2 Antimicrobial Activities of Linear Decapeptides
(Source: ref. 2. Reproduced by kind permission of Kyoritsu Shuppan Co.)

Peptide†1	Minimum inhibitory concentration (μg/m*l*)	
	B. subtilis	*S. aureus*
Gramicidin S	0.8	1.6
HCO–(D-Phe–Pro–Val–Orn–Leu)$_2$–EA (**4.4**)	50	100
HCO–(Val–Orn–Leu–D-Phe–Pro)$_2$–EA (**4.5**)	6.3	25
H–( )$_2$–OH (**4.1**)†2	3.1	25

†1 All optically active amino acids are of the L-configuration unless otherwise indicated. EA, $NHCH_2CH_2OH$.
†2 The compound (**4.1**) has the same amino acid sequence as the compound (**4.5**).

It is interesting to note that not only oligopeptides but also some special amino acid derivatives exhibit considerable antibacterial activity. Nakamiya *et al.*[178)] prepared a series of DL-amino acid alkyl esters with C_{10} to C_{18} alkyl groups and tested their antibacterial activity. As shown in Table 4.3, DL-lysine dodecyl ester (**4.6**) was most effective.

TABLE 4.3 Antimicrobial Activities of Amino Acid Alkyl Ester and a Related Compound

Compound	Minimum inhibitory concentration (μg/ml)	
	S. aureus	*E. coli*
Dodecanoyl-L-Lys	> 400	> 400
DL-Lys dodecyl ester (**4.6**)	6.3	12.5

Ruttenberg *et al.* prepared open-chain tyrocidin B[179] by selective cleavage of the D-phenylalanylprolyl bond, and reported that the resulting peptide was inactive.[167] It is necessary to investigate the antibacterial activity of such open-chain analogs of tyrocidines in more detail, since the open-chain analogs (e.g., **4.1** and **4.5**) prepared by Erlanger *et al.* and Makisumi *et al.* were active. Linear gramicidins (Fig. 2.2 in Chapter 2) have a structure similar to that of the weakly active formyl peptide ethanolamide (**4.4**) in appearance, but exhibit strong antibacterial activity. Recently, several active shortened analogs of gramicidin A have been synthesized by Miroshnikov *et al.*[180] and Srinivasa and Ramachandran.[181]

4.1.2 LINEAR POLYPEPTIDES AND RELATED CYCLIC PEPTIDES

Matsuura *et al.*[119, 120] synthesized linear polypeptides (**4.8** and **4.8**) related to gramicidin S as well as the corresponding macro-ring analogs of gramicidin S (**4.10** and **4.11**) (Table 4.4). The linear pentapeptide (**4.7**) was inactive,

TABLE 4.4 Antimicrobial Activities of Linear Oligopeptides Related to Gramicidin S and Macro-ring Analogs of Gramicidin S (Source: ref. 2. Reproduced by kind permission of Kyoritsu Shuppan Co.)

Peptide†1	Minimum inhibitory concentration (μg/ml)	
	B. subtilis	*S. aureus*
H–(Val–Orn–Leu–D-Phe–Pro)–OH (**4.7**)	>100	>100
H–( )$_2$–OH†2 (**4.1**)	50	50
H–( )$_3$–OH†2 (**4.8**)	50	50
H–( )$_4$–OH†2 (**4.9**)	10	20
SemiGS (cyclic pentapeptide)	>100	>100
GS (cyclic decapeptide)	2	5
SesquiGS (cyclic pentadecapeptide) (**4.10**)	20	50
DiGS (cyclic eicosapeptide) (**4.11**)	10	20

†1 All optically active amino acids are of the L-configuration unless otherwise indicated. GS, gramicidin S.

†2 The compounds have the same repeated pentapeptide sequence as the linear pentapeptide (**4.7**).

as was the corresponding semigramicidin S (a cyclic pentapeptide). As shown in Table 4.4, the specific activity of each linear peptide (**4.8** or **4.9**) was approximately the same as that of the corresponding macro-ring analog (**4.10** or **4.11**) of gramicidin S. Both the linear peptide (**4.8** or **4.9**) and the corresponding macro-ring analog (**4.10** or **4.11**) may have a spatial structure of the type **3.44** or **3.45** described in Section 3.2.1. Furthermore, the level of specific activity increased somewhat with an increase of molecular size (from pentadecapeptide to eicosapeptide) in both the linear and macro-ring analogs.

4.1.3 POLY-α-AMINO ACIDS RELATED TO GRAMICIDIN S

In 1953, Katchalski *et al.*[182] reported that some basic polyamino acids such as poly-DL-ornithine exhibit strong antibacterial activities, whereas neutral or acidic polyamino acids are inactive, and discussed the mode of action of polyamino acids on bacteria. They also synthesized a variety of copolymers composed of the constituent amino acids of gramicidin S and other amino acids such as alanine.[183,184] The antibacterial activities of these polymers and copolymers are summarized in Table 4.5.

The random copolymer (**4.12**) composed of the constituent amino acids of gramicidin S in a suitable ratio inhibits the growth of the microorganisms tested at nearly the same concentration as the natural antibiotic. An alteration in the configuration of phenylalanine from D to L produced little change in the activity. The elimination of L-proline or its replacement by sarcosine resulted in no loss of the activity. A copolymer composed of equimolar a-

TABLE 4.5 Antimicrobial Activities of Amino Acid Polymers and Copolymers Related to Gramicidin S (Source: ref. 2. Reproduced by kind permission of Kyoritsu Shuppan Co.)

Amino acid composition in polymer or copolymer†1	DP†2	Minimum inhibitory concentration (μg/ml)	
		E. coli	*M. pyogenes*
Gramicidin S		2.5	0.5
Val Orn Leu D-Phe Pro (**4.12**)	30	2.5	2.5
Val Orn Leu Phe Pro	40	10	10
Val Lys Leu D-Phe Pro	40	2.5	2.5
Val Orn Leu D-Phe Sar	25	5	5
Orn Leu DL-Ala	45	2.5	2.5
Orn Leu (1:1)	23	0.3	2.5
Leu DL-Ala (1:3)	26	>250	>250
poly-Orn	73	1.5	1.5

†1 All optically active amino acids are of the L-configuration unless otherwise indicated.
†2 Average number of amino acid residues per molecule.

mounts of L-ornithine and L-leucine is strongly active, even surpassing natural gramicidin S in its activity against *Escherichia coli.* These results reveal the importance of basic amino acids and of amino acids with a hydrophobic side chain for the activity of gramicidin S. In studies on the nature of the antibacterial action, it was shown that gramicidin S and the copolymer (**4.12**) had bactericidal activity, while poly-L-ornithine was found to be considerably less bactericidal, though it did inhibit the growth of bacteria.

The marked similarity in the modes of antibacterial action of gramicidin S and the copolymer (**4.12**) suggests the existence of a secondary structure such as **4.13** in the polymer. There is some possibility of incorporation of cyclic compounds into the polyamino acids because the polyamino acids were synthesized by the *N*-carboxylic anhydride method. The level of specific acti-

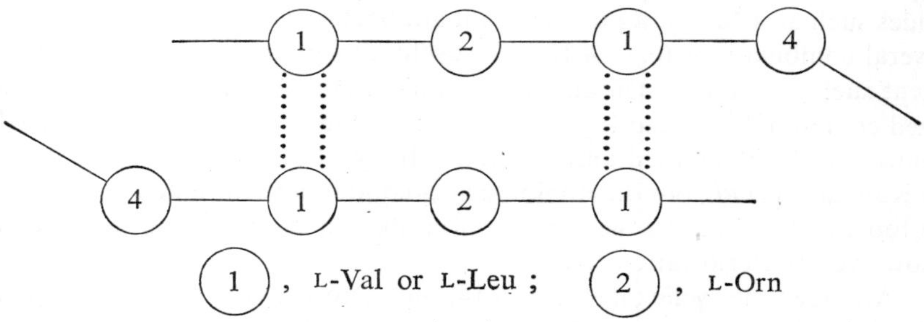

(4.13)

vity of the copolymers is the same as that of gramicidin S, whereas the specific activity of the macro-ring analogs, sesquigramicidin S and digramicidin S (**4.10** and **4.11**), is about one-fifth to one-tenth (Table 4.4). It is still difficult to explain the difference between the specific activities of the polyamino acids and the macro-ring analogs related to gramicidin S at this time.

4.2 Peptides with Smaller or Larger Ring Structures Compared with Gramicidin S

4.2.1 PEPTIDES WITH SMALLER RING STRUCTURES

We undertook systematic studies on analogs of gramicidin S involving the ring structure of gramicidin S. Early in 1960, we emphasized that some peptide antibiotics such as gramicidin S, tyrocidines and polymyxins whose structures had been elucidated or assumed at that time possessed the following structural features in common. (1) A cyclic structure, (2) one or more D-amino acids, and (3) one or more basic amino acids. Our initial studies were started with these features in mind.

As the first target peptide with the simplest structure possessing the characteristics mentioned above, we selected a cyclic dipeptide (diketopiperazine) containing dipeptide segments around the ornithine residues in gramicidin S. However, the cyclic dipeptide (**4.14**) and its 12 analogs had no activity even at a concentration of 100 μg/ml. In 1964 we reported these results in the first

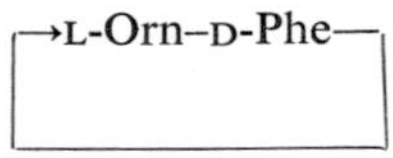

(4.14)

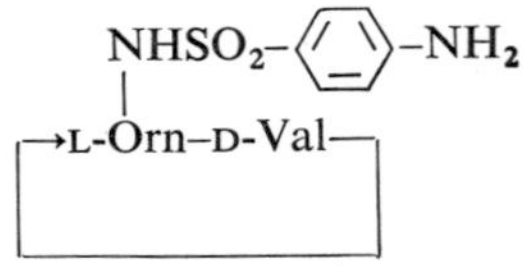

(4.15)

paper[185)] of a series on peptide antibiotics. Postulating that certain cyclopeptides such as **4.14** might possess latent antibacterial activity, we synthesized several combinations (e.g., **4.15**) of a cyclic dipeptide with an antibacterial agent such as *p*-amino-benzenesulfonylamide. However none of the synthesized compounds showed higher activity than the parent sulfonamide.[186)] In contrast with the marked antibacterial activity of poly-L-ornithine reported by Katchalski *et al.*, our diketopiperazines containing ornithine or lysine may be too small to bind to the surface of gramicidin S-sensitive bacteria or to show even bacteriostatic activity.

Although the cyclic dipeptide (**4.14**) shows no activity, several groups of diketopiperazine antibiotics have been reported in nature, including gliotoxins,[187–9)] sporidesmins,[189,190)] aranotins[191,192)] and albonoursin.[193,194)] These naturally occurring diketopiperazine derivatives contain an unusual amino acid unit and some of them show antiviral or antitumor activity as well as antibacterial activity.

We next planned to synthesize other types of cyclic peptides with somewhat larger ring size than diketopiperazine and with the partial amino acid sequences found in gramicidin S and tyrocidines, because we considered that structural features such as an appropriate ring size and/or a specific amino acid sequence might be factors governing the appearance of the antibacterial activity of gramicidin S. In the course of studies on the cyclization of a linear pentapeptide active ester, Waki and Izumiya noted the formation of semigramicidin S (a cyclic pentapeptide) (**4.16**) ($n = 0$).[17,18)] However, semigrami-

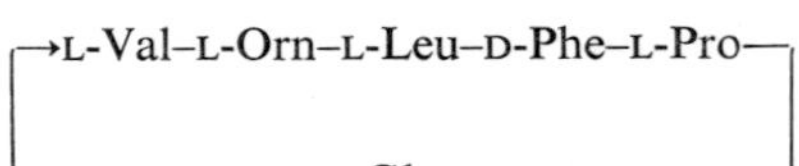

(4.16)

cidin S was inactive (see Table 4.4). The conformation of semigramicidin S is quite different from that of gramicidin S or its active analogs; this will be further discussed in Sections 5.2 and 5.3. Eighteen cyclic pentapeptides related to gramicidin S have been synthesized in our laboratory, and three in Rothe's laboratory, as shown in Tables 3.4 and 3.5. None of these cyclic pentapeptides showed activity. The cyclic hexapeptide (**4.16**) ($n = 1$) still showed no activity,[195] and even the cyclic heptapeptide (**4.16**) ($n = 2$)[196] showed no activity against any of the microorganisms tested.

The existence of active fragments or an active core in a molecule is well known in the field of peptide hormones. For example, linear active fragments of adrenocorticotropin can easily bind to a receptor site because of the flexibility of the linear peptide. However, a cyclic peptide (**4.16**) ($n = 0$, 1 or 2) containing a possible active fragment (e.g., the pentapeptide sequence from positions 1 to 5 of gramicidin S) might not bind or fit to the receptor sites of gramicidin S-sensitive bacteria because of an unfavorable oligopeptidyl ring structure (**4.16**).

4.2.2 PEPTIDES WITH LARGER RING STRUCTURES

A typical macro-ring analog of gramicidin S (30-membered ring) is sesquigramicidin S (45-membered) or digramicidin S (60-membered), the antibacterial activity of which is shown in Table 4.4. A discussion of their activities and structures in comparison with those of the corresponding linear oligopeptides is also given in Section 4.1.2. [βAla$^{5,5'}$]-gramicidin S[92] is another analog with a 32-membered ring by virtue of the two additional methylene groups. Recently Takiguchi and Izumiya[197] synthesized isogramicidin S (36-membered), which can also be regarded as a macro-ring analog, as can be seen from the partial structures of isogramicidin S (**4.17**) and gramicidin S (**4.18**). Both isogramicidin S and [βAla$^{5,5'}$]-gramicidin S showed no antibac-

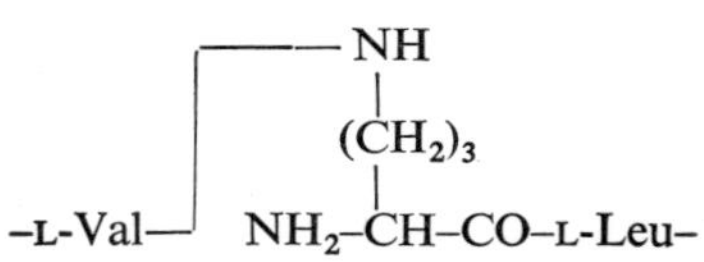

(**4.17**)

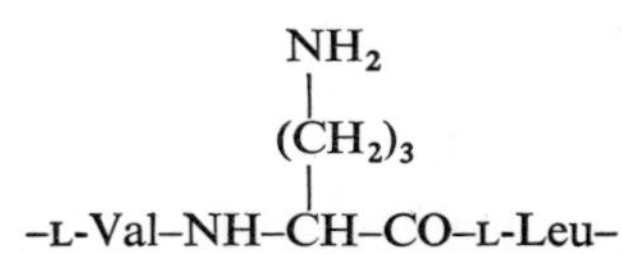

(**4.18**)

terial activity. The ORD curve of the former was similar to that of gramicidin S, whereas that of the latter was quite different. It seems to be rather difficult to explain the absence of activity of these two analogs simply on the basis of the ORD results obtained.

4.3 Substituted Analogs of Gramicidin S

In 1958, two important initial studies were reported. One was the synthesis of [L-Lys$^{2,2'}$]-gramicidin S by Schwyzer and Sieber[64,198] and the other was the modification of natural gramicidin S by Stepanov *et al.*[150] Since then, large numbers of such analogs of gramicidin S have been synthesized and their antibacterial activities have been examined. The synthesized amino acid replacement analogs and the modified analogs derived from natural gramicidin S that have been prepared so far are summarized in Tables 4.6 and 4.7, respectively, together with their antibacterial activities. Most of the amino acid replacement analogs have been synthesized in our laboratory.

TABLE 4.6 Antimicrobial Activities of Synthetic Amino Acid Replacement Analogs of Gramicidin S

No.	Amino acid sequence of cyclic decapeptide†1						Minimum inhibitory concentration (μg/ml)†2		Ref.
							B. subtilis	*S. aureus*	
1	Val$^{1,1'}$	Orn$^{2,2'}$	Leu$^{3,3'}$	D-Phe$^{4,4'}$	Pro$^{5,5'}$	(GS)	5	5	17,18
2	Gly$^{1,1'}$						100	100	106
3	Gly$^{1,1'}$				Gly$^{5,5'}$		>100	>100	88
4	Ala$^{1,1'}$						5	5	106
5	Ala$^{1,1'}$		Ala$^{3,3'}$				>100	>100	126
6	Ser$^{1,1'}$		Ser$^{3,3'}$				no activity		84
7	Val^{1}Cys(MBzl)$^{1'}$						10	10	133
8	Val^{1}Cys(MBzl)$^{1'}$		Cys(MBzl)^{3}Leu$^{3'}$				10	40	199
9	Val^{1}Cys$^{1'}$		Cys^{3}Leu$^{3'}$				10–20	40–80	199
10	Val^{1}MeVal$^{1'}$						80	40	123
11	MeVal$^{1,1'}$						160	80	123
12	Hmb$^{1,1'}$						5–20	45–90	200,201
13	Hmb$^{1,1'}$		Hmp$^{3,3'}$				> 500	> 500	200,201
14	Leu$^{1,1'}$						5	10	107
15	Phe$^{1,1'}$		Phe$^{3,3'}$				80	>160	127
16		Dbu$^{2,2'}$					5	10	108
17		Lys$^{2,2'}$					10	10	108,198
18		Orn^{2} His$^{2'}$					5	5	131
19		His$^{2,2'}$					40	80	131
20		Lys$^{2,2'}$		Gly$^{4,4'}$				>200	135
21			Gly$^{3,3'}$				>100	>100	109
22			Ala$^{3,3'}$				20	50	109
23			Val$^{3,3'}$				25	50	124
24			Leu^{3}MeLeu$^{3'}$				5	5	121,122
25			MeLeu$^{3,3'}$				5	5	121,122
26				Gly$^{4,4'}$			50	>100	82,100
27				Aib$^{4,4'}$			>100	>100	82
28				Ala$^{4,4'}$			>100	>100	82

(continued)

TABLE 4.6—*Continued*

No.	Amino acid sequence of cyclic decapeptide†[1]	Minimum inhibitory concentration (μg/ml)†[2] *B. subtilis*	*S. aureus*	Ref.
29	D-Ala$^{4,4'}$	40	40	82,111
30	D-Val$^{4,4'}$	10	10	112
31	D-Leu$^{4,4'}$	5	10	112
32	Phe$^{4,4'}$	45–180	300–700	90,135 201,202
33	Pro(4F)$^{5,5'}$	2.5		87
34	Pro^{5}Gly$^{5'}$	10	10	89
35	Gly$^{5,5'}$	10	10	67,77, 78,83, 90,129, 137,143, 201,202
36	Sar$^{5,5'}$	5	10	91
37	βAla$^{5,5'}$	160	80	92
38	Aib$^{5,5'}$	>100	>100	117
39	Leu$^{5,5'}$	5	10	134
40	Phe$^{5,5'}$	10	10	130
41	δ-Ava$^{5-1'}$	20	40	94
42	δ-Ava$^{5-1',5'-1}$	>100	>100	94
43	δ-Ava$^{4-5}$	5	10	118,203
44	δ-Ava$^{4-5,4'-5'}$	>100	>100	93
45	[Cys^1]-GS [Cys^1]-GS	>100	>100	133
46	IsoGS	>100	>100	197
47	Iso-[Lys$^{2,2'}$]-GS	>100	>100	197
48	RetroGS	50	50	90,113, 114,201, 202
49	EnantioGS	5	5	90,201, 202
50	RetroenantioGS	90–180	300–750	90
51	Retro-[Gly$^{5,5'}$]-GS	60–180	>500	90,146, 147,201, 202
52	Enantio-[Gly$^{5,5'}$]-GS	45–180	90–180	90,201, 202
53	Retroenantio-[Gly$^{5,5'}$]-GS	60–180	>500	90,146, 147

†[1] Only the varied residues are shown except in the case of the first compound. All optically

active amino acids are of the L-configuration unless otherwise indicated. GS, gramicidin S; MBzl, *p*-methoxybenzyl; Hmb, 2-hydroxy-3-methylbutanoic acid; Hmp, 2-hydroxy-4-methylpentanoic acid; Dbu, α,γ-diaminobutyric acid; Aib, α-aminoisobutyric acid; Pro (4F), 4-fluoroproline; δ-Ava, δ-aminovaleric acid.

†2 In certain cases the figures are derived by calculation on the basis of those for gramicidin S as a standard.

TABLE 4.7 Antimicrobial Activities of Modified Analogs of Gramicidin S

No.	Amino acid sequence of cyclic decapeptide†1	Minimum inhibitory concentration ($\mu g/ml$)†2			Ref.
		B. Subtilis	*S. aureus*	Other micro-organisms	
1	$Val^{1,1'}$ $Orn^{2,2'}$ $Leu^{3,3}$ D-$Phe^{4,4'}$ $Pro^{5,5'}$ (GS)	5	5		17, 18
2	Orn^{2} $Orn(Ac)^{2'}$	80	40		†3
3	$Orn(Ac)^{2,2'}$			no activity	204–206
4	$Orn^{2}Orn(hexanoyl)^{2'}$	80	>160		207, †3
5	$Orn(nonanoyl)^{2,2'}$	10			207
6	$Orn^{2}Orn(decanoyl)^{2'}$	5			207
7	$Orn(decanoyl)^{2,2'}$	5			207
8	$Orn(undecanoyl)^{2,2'}$	2.5			207
9	$Orn^{2}Orn(dodecanoyl)^{2'}$	5			207
10	$Orn(dodecanoyl)^{2,2'}$	5			207
11	$Orn(tridecanoyl)^{2,2'}$	10			207
12	$Orn^{2}Orn(tetradecanoyl)^{2'}$	10			207
13	$Orn(tetradecanoyl)^{2,2'}$	40			207
14	$Orn(hexadecanoyl)^{2,2'}$	160			207
15	$Orn(Bz)^{2,2'}$			no activity	204, 205
16	$Orn(Pht)^{2,2'}$			500	205, 208
17	$Orn^{2}Orn(Z)^{2'}$	40	10		†3
18	$Orn(Z)^{2,2'}$			500	205
19	$Orn^{2}Orn(Dns)^{2'}$		50		209
20	$Orn(Dns)^{2,2'}$		>400		209
21	$Orn(Cbm)^{2,2'}$			500	205, 208
22	$Orn(Gly)^{2,2'}$			20	150, 205 206
23	$Orn(6\text{-aminohexanoyl})^{2,2'}$			40	150, 205
24	$Orn(9\text{-aminononanoyl})^{2,2'}$			40	150, 205
25	$Orn(Dns\text{-}Orn)^{2,2'}$			25	206
26	$Orn(Suc)^{2,2'}$			500	205, 208
27	$Orn^{2}Orn(Cm)^{2'}$		very low activity		210

(continued)

TABLE 4.7—*Continued*

No.	Amino acid sequence of cyclic decapeptide†1		Minimum inhibitory concentration (μg/ml)†2 B. subtilis	*S. aureus*	Other micro-organisms	Ref.
28	$Orn(Cm)^{2,2'}$			no activity		205, 210
29	$Orn(Me_2)^{2,2'}$		5		5	204, 205
30	$Orn(Et_2)^{2,2'}$		5		5	204, 205
31	$Orn(Me_2C_nH_{2n+1})^{2,2'}$ (n = 2,4,6,8,10,12,14,16,18)		10			211
32	$Arg^{2,2'}$		5	5	5	151, 205, 208, 212
33	$Arg(NO_2)^{2,2'}$				no activity	205
34	$Nva(4OH)^{2,2'}$				no activity	205, 212
35	$Glu^{2,2'}$		400			212
36		D-$Phe(NO_2)^{4,4'}$		5		213
37		D-$Phe(NH_2)^{4,4'}$		75		213, 214
38		D-$Phe(I)^{4,4'}$		40		214
39		D-$Phe(Cl)^{4,4'}$	no activity			215
40		D-$Cha^{4,4'}$	20	10		82
41	$Arg(NO_2)^{2,2'}$	D-$Phe(NO_2)^{4,4'}$		75		213
42	$Arg(NO_2)^{2,2'}$	D-$Phe(NH_2)^{4,4'}$		75		213

†1 See the footnote (†1) in Table 4.6. Ac, acetyl; B_z; benzoyl; Dns, dansyl; Cbm, carbamoyl; Suc, succinyl; Cm, carboxymethyl; Nva(4OH), 4-hydroxynorvaline; Cha, cyclohexylalanine.

†2 See the footnote (†2) in Table 4.6.

†3 The compound was synthesized by Minari *et al.* in our laboratory.

4.3.1 FUNCTION OF EACH AMINO ACID RESIDUE IN GRAMICIDIN S

The following discussion on the role of each amino acid residue in gramicidin S in connection with its biological activity is based on the summarized data in Tables 4.6 and 4.7.

A. Function of the Amino Acid Residues at Positions 1 and 1′

The replacement of valine residues in gramicidin S by L-alanine or L-leucine residues does not affect the specific activity (No. 4 or 14 in Table 4.6). The striking feature is the sharp drop of the activity upon glycine substitution (No. 2 in Table 4.6). One of the valine residues in gramicidin S is replaceable by *S*-*p*-methoxybenzyl-L-cysteine without any significant effect on the antibacterial activity (No. 7 in Table 4.6).

In Fig. 4.1 we assume that the bulkiness or hydrophobicity of the side chains of four amino acid residues located on the upper side of the peptide backbone is necessary for the interaction between gramicidin S and sensitive bacteria. Thus, side chains as bulky or hydrophobic as methyls, at least, may be required at positions 1 and 1′ for the interaction between gramicidin S and sensitive bacteria. The features of some interesting analogs of gramicidin S containing *N*-methylamino acid (Chapter 5), α-hydroxy acid or δ-aminovaleric acid residues at poitions 1 and/or 1′ and a dimerized analog, *S,S′*-bi ([L-hemicystine^1]-gramicidin S) (Chapter 5), will be described in a separate section.

B. Function of the Amino Acid Residues at Positions 2 and 2′

Three methylene groups of ornithine side chains in gramicidin S can be replaced by either two methylene groups ([L-α,γ-diaminobutyric acid$^{2,2'}$]-analog) or four ([L-Lys$^{2,2'}$]-analog) without loss of biological activity (No. 16 or 17 in Table 4.6). Thus, strict positioning of the cationic moiety ($-\overset{+}{N}H_3$) in gramicidin S is not necessary for the activity, and the receptor pocket of gramicidin S-sensitive bacteria may be fairly loose (see Fig. 4.1). Diaminoacylated analogs (Nos. 22–24 in Table 4.7) of gramicidin S in which the aminoacyls are glycine, 6-aminohexanoyl and 9-aminononanoyl show lower activities than the parent antibiotic, and the activities decrease with increasing chain length of the aminoacyl residues.

Single replacement of ornithine residues in gramicidin S by L-histidine, with its weakly basic imidazolyl group, affords a fully active [L-His^2]-analog (**4.19**) (No. 18 in Table 4.6). In contrast, di-substituted [L-His$^{2,2'}$]-gramicidin S (**4.20**) (No. 19 in Table 4.6) is only 1/8 or 1/16 as active as gramicidin S. We can tentatively explain the apparent discrepancy as follows: the mono-substituted analog (**4.19**) can gain access to the bacterial membrane by ionic interaction between the amino group and phospholipid.[206)] A weakly basic imidazolyl group also can make a salt-like linkage with the phospholipid, when the mono-substituted analog (**4.19**) approaches the membrane. However, ionic interaction between the di-substituted analog (**4.20**) and the membrane may be so weak that it is difficult for **4.20** to come close to the lipid membrane, thus weakening its antimicrobial activity. Histidine is known to constitute a part of the catalytic site in several esterolytic enzymes, such as α-chymotrypsin and subtilisin. It is interesting to note that these analogs (**4.19** and **4.20**) with histidine showed esterolytic activity against *p*-nitrophenyl acetate comparable to or greater than that of imidazole.[131)]

A structural isomer of gramicidin S, isogramicidin S (partial structure, **4.17**) shows no antibacterial activity (No. 46 in Table 4.6). We assume the reasons for this inactivity of isogramicidin S to be as follows: conformational change due to the enlargement of the ring frame of gramicidin S or loss of the interaction between isogramicidin S and the bacteria because of an excessive distance between anionic portions of the bacteria (see Fig. 4.1) and cationic amino groups (α-amino groups of ornithine residues) directly connected with the ring frame of the analog. To investigate the former hypothesis, the use of D-ornithine may be informative. To study the latter hyphothesis, it would be interesting to replace the basic ornithine residues of gramicidin S with L-diaminopropionic acid or diaminoacetic acid, with a shorter side chain than ornithine or L-α,γ-diaminobutyric acid.

Table 4.7 shows a number of ornithine-modified analogs of gramicidin S that have been prepared. This is largely because of the specific function of ornithine residues with their basic character, and also because of the reactive nature of the ornithine functional group compared with other amino acid residues in gramicidin S. Alkylation or guanidination of the δ-amino groups of ornithines in gramicidin S affords compounds (Nos. 29–32 in Table 4.7) with strongly basic character, which retain full activity. Diacetyl-gramicidin S (No. 3 in Table 4.7) has a conformation similar to that of gramicidin S, but no activity. Unexpectedly, other diacylated neutral analogs (Nos. 5, 7, 8 etc., in Table 4.7) show substantial antibacterial activity. The activities of these diacylated analogs were found to increase with increase in the number of carbon atoms of acyl groups, reaching a maximum at around C_{10} to C_{12} and decreasing thereafter. We cannot explain the wide variation in the activities of diacetyl-gramicidin S and other diacylated analogs at present.

Monoacylated analogs of gramicidin S with one amino group (e.g., No. 2 in Table 4.7) have a conformation similar to that of gramicidin S, and retain some activity. This may be a result of weak interaction between the monoacylated analogs and anionic moieties in the bacteria (see Fig. 4.1). The replacement of ornithine residues by L-histidine residues, which have a weakly basic character, lowers the activity, reflecting a similar weak interaction of the derivative (No. 19 in Table 4.6) and the bacteria. On the other hand, changing the basic character of gramicidin S to neutral (Nos. 15, 16, 18, 27, 33, 34, 41 and 42 in Table 4.7) or acidic (Nos. 26, 28 and 35 in Table 4.7) results in a loss of activity. These results suggest that the basic character of ornithine residues at positions 2 and 2′ plays an important role in the activity of gramicidin S.

C. Function of the Amino Acid Residues at Positions 1,1′ and 3,3′

The replacement of amino acid residues at positions 1,1′ and 3,3′ by hydrophilic amino acids such as L-serine affords an inactive compound (No. 6 in Table 4.6). The same compound was prepared by the solid-phase method.[216] Replacement by aliphatic amino acids such as L-alanine also results in a loss of the activity (No. 5 in Table 4.6). However, the analog (No. 15 in Table 4.6) with an aromatic amino acid, L-phenylalanine, in these positions retains activity. Simultaneous replacement of one valine and one leucine residues in gramicidin S by *S-p*-methoxybenzyl-L-cysteine or half-L-cystine resulted in a slight loss of activity (No. 8 or 9 in Table 4.6). Thus the side chains of the four amino acid residues at these positions must be more hydrophobic or larger than methyl groups for effective interaction between the analogs and sensitive bacteria (see Fig. 4.1).

D. Function of the Amino Acid Residues at Positions 3 and 3′

Increasing bulkiness of the side chains of amino acid residues at positions 3 and 3′ with glycine, L-alanine and L-leucine (in natural gramicidin S) increases the specific activity stepwise (Nos. 21, 22 and 1 in Table 4.6). However, the analog (No. 23 in Table 4.6) having bulky L-valines at these positions shows relatively weaker activity. Although there are some qualitative differences in the effects of the amino acid side chains at positions 3,3′ and also 1,1′ in gramicidin S, the activity seems to be more sensitive to 3,3′ substitution than to 1,1′ substitution. These differences cannot be explained by the working hypothesis shown in Fig. 4.1.

E. Function of the Amino Acid Residues at Positions 4 and 4′

Amino acid residues at positions 4 and 4′ also play a specific role in the biological activity, like the amino acid residues at positions 2 and 2′. Increase of the activity from glycine, D-alanine, D-valine to D-leucine substitution at these positions is similar to the increase seen on amino acid substitution at positions 3 and 3′ (Nos. 26, 29–31 in Table 4.6). Therefore hydrophobicity of the side chains at positions 4 and 4′ appears to increase the activity, though this should not be the case, because these side chains should be directed to the hydrophilic side of the molecule (Fig. 3.3). Chandrasekaran *et al.*[217)] showed that the most stable β-turn geometry contains one D-amino acid residue, and Venkatachalam[218)] suggested that larger side chain groups may reduce the flexibility. We assume that the bulkier the side chains in the D-amino acid residues are, the more stable the β-turn becomes, and this results in high activity.

On the other hand, [α-aminoisobutyric acid$^{4,4'}$]-, [L-Ala$^{4,4'}$]-gramicidin S and all-L-gramicidin S ([L-Phe$^{4,4'}$]-analog) have no activity (Nos. 27, 28 and 32 in Table 4.6). Kawai and Nagai[82)] suggested that the conformations of these inactive analogs are quite different from the stable β-sheet conformation of gramicidin S or its active analogs on the basis of ORD and CD results.

Using our working hypothesis (Fig. 4.1), it may be possible to explain the degree of activity of analogs of gramicidin S (e.g., Nos. 4 and 14 in Table 4.6) having a conformation similar to that of natural gramicidin S, but it is not straightforward to predict the activity of analogs (e.g., all-L-gramicidin S) having different conformations. Aromaticity of amino acid residues at positions 4 and 4′ is not necessary for activity because hydrogenated gramicidin S (No. 40 in Table 4.7) still retains some activity.

F. Function of the Amino Acid Residues at Positions 5 and 5′

[Gly$^{5,5'}$]-gramicidin S (No. 35 in Table 4.6), synthesized for the first time in our laboratory, has been used as a model peptide for various purposes because of its high activity and favorable conformation, as mentioned in the preceding chapter. Recently El-Maghraby[87)] reported the synthesis of an analog (No. 33 in Table 4.6) more potent than natural gramicidin S by the replacement of proline residues at positions 5 and 5′ with L-4-fluoro-proline residues. It is usual to replace proline with some other imino acid such as sarcosine to obtain a proline-related analog in the field of biologically active peptides. Sarcosine substitution of proline residues at positions 5 and 5′ in gramicidin S affords a compound (No. 36 in Table 4.6) which is as active as gramicidin S.

Recent development of peptide cyclization methods using the practically racemization-free *N*-hydroxysuccinimide active ester or azide coupling reaction makes it possible to place optically active amino acids at positions 5 and 5′ of gramicidin S. It is worth noting that the recently synthesized compounds [L-Leu$^{5,5'}$]- and [L-Phe$^{5,5'}$]-gramicidin S (Nos. 39 and 40 in Table 4.6) also show strong activities. Thus it is possible to replace prolines at positions 5 and 5′ by a variety of amino acids (sarcosine, glycine, phenylalanine, etc.) without loss of activity. If we assume that the analogs at positions 5 and 5′ in gramicidin S have a conformation similar to that of gramicidin S, amino acid residues at positions 5 and 5′ may not have much effect on the interaction with gramicidin S-sensitive bacteria (see Fig. 4.1). The inactivity of the analogs containing an unusual amino acid, such as β-alanine or α-aminoisobutyric acid, may reflect disorder of the gramicidin S-like conformation by the unusual amino acid residues in the analogs (Nos. 37 and 38 in Table 4.6).

G. Relationship between Primary Structure and Activity

The relationships between the structure and activity of gramicidin S mentioned in this section can be briefly summarized as follows. (1) All analogs with hydrophobic side chains in the positions 1, 3, 1′ and 3′ are active, and the degree of activity is related to the hydrophobicity of the side chains. (2) In the positions 2 and 2′, at least one charged group with a more basic character than the amino group seems to be indispensable for biological activity. (3) D-Configuration in the positions 4 and 4′ is important for the activity, but aromatic nature in these positions is not essential. (4) A variety of amino acids except for unusual amino acids such as β-alanine or α-aminoisobutyric acid are exchangeable at positions 5 and 5′ without loss of activity.

All these results can be well explained by the "sidedness" hypothesis[65,219] that two hydrophilic side chains are on one side of the molecule and four hydrophobic ones are on the other side (see also Fig. 3.3 and Fig. 4.1). This will be described in detail in a separate chapter.

4.3.2 ANALOGS CONTAINING δ-AMINOVALERIC ACID

Muramatsu and Sōfuku attempted to elucidate the significance of intramolecular hydrogen bondings in gramicidin S by the following approach. Two consecutive amino acid residues of gramicidin S are replaced by δ-aminovaleric acid, and consequently one amide group (–CONH– or $-CO-\overset{|}{N}-$) is replaced by one ethylene group ($-CH_2CH_2-$), which has almost the same length as –CONH– but does not participate in hydrogen bonding. This approach makes it possible to clarify the contributions of peptide bonds and hydrogen bonds of gramicidin S to the antibacterial activity of the antibiotic.

Sōfuku[118] synthesized [δ-aminovaleric acid$^{4-5}$]-gramicidin S (No. 43 in Table 4.6), in which one D-phenylalanyl-L-prolyl moiety of the natural peptide was replaced with a δ-aminovaleric acid residue. The analog (No. 43 in Table 4.6) showed ORD and CD spectra similar to those of gramicidin S and had a fairly high activity.[118,203] These findings suggest that not every peptide bond is indispensable to the function and maintenance of the backbone structure of gramicidin S. The structural formula of the analog (**4.21**), showing the hydrogen bonds, illustrates why the activity of this analog is

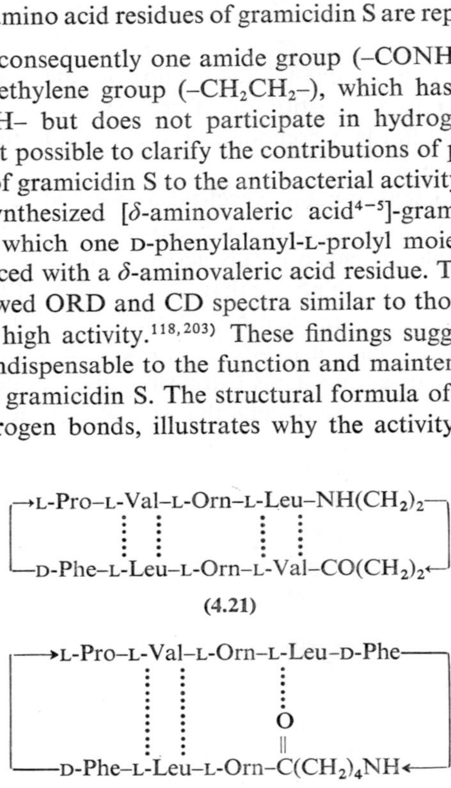

(4.21)

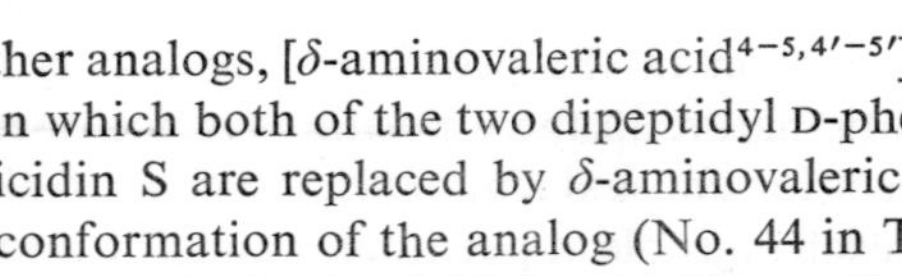

(4.22)

maintained. Another analogs, [δ-aminovaleric acid$^{4-5,4'-5'}$]-gramicidin S (No 44 in Table 4.6), in which both of the two dipeptidyl D-phenylalanyl-L-prolyl residues of gramicidin S are replaced by δ-aminovaleric acids, showed no activity, and the conformation of the analog (No. 44 in Table 4.6) was presumed to be random on the basis of ORD and CD studies.[93]

[δ-Aminovaleric acid$^{5-1'}$]-gramicidin S (**4.22**) (No. 41 in Table 4.6), in which one L-prolyl-L-valyl residue is replaced by δ-aminovaleric acid, showed a similar ORD spectrum and some antibacterial activity.[94] On the basis of a CD study they assumed that the compound (**4.22**) has a partially flexible structure compared with the rigid structure of gramicidin S, mainly because of the lack of one amide hydrogen between L-proline and L-valine residues.[94] [δ-Aminovaleric acid$^{5-1', 5'-1}$]-gramicidin S, which lacks two amide hydrogens, was inactive.[94]

4.3.3 DEPSIPEPTIDE ANALOGS CONTAINING AN α-HYDROXY ACID

A Russian group[85,200] synthesized two depsipeptide analogs, [L-2-hydroxy-3-methylbutanoic acid$^{1,1'}$]-gramicidin S (**4.23**) and [L-2-hydroxy-3-methylbutanoic acid$^{1,1'}$, L-2-hydroxy-4-methylpentanoic acid$^{3,3'}$]-gramicidin S (**4.24**), containing α-hydroxy acids instead of α-amino acids. This is another interesting approach to elucidate the contribution of the intramolecular hydrogen bonds in gramicidin S to its conformational stability and biological

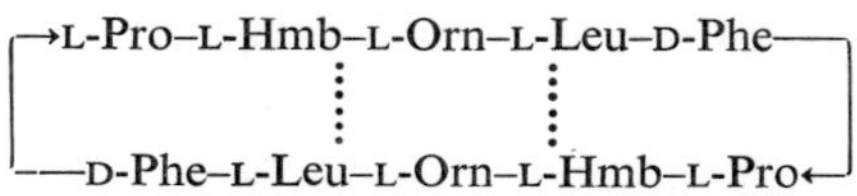

Hmb, 2-hydroxy-3-methylbutanoic acid

(4.23)

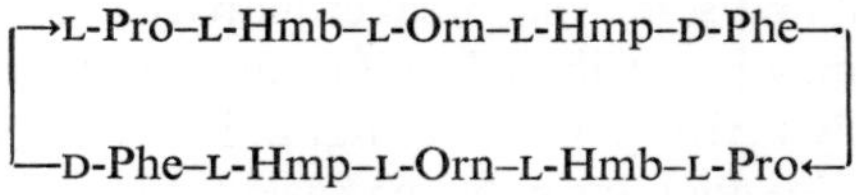

Hmp, 2-hydroxy-4-methylpentanoic acid

(4.24)

activity. In such a depsipeptide analog, an amide bond (–CONH–) is replaced by an ester bond $(-\overset{\overset{\displaystyle O}{\|}}{C}-O-)$ to eliminate the possibility of hydrogen bonding by an amide proton. Thus the analog (**4.23**) has only two and the analog (**4.24**) has no intramolecular hydrogen bonds, although gramicidin S has four. The former analog (**4.23**) had some antibacterial activity, while the latter (**4.24**) was inactive, with less conformational stability than gramicidin S.[200,201] These results show that the rigid spatial structure of gramicidin S stabilized by intramolecular hydrogen bonds is required for manifestation of the biological activity of gramicidin S.

4.4 Tyrocidines and Their Analogs

The antibacterial activity of tyrocidines has not yet been determined quantitatively. Recently Okamoto *et al.*[164] reported a quantitative assay of the activity of tyrocidines and several tyrocidine A analogs (see Fig. 2.1). As shown in Table 4.8, there are only small differences in the specific activities

TABLE 4.8 Antimicrobial Activities of Cyclic Decapeptides Related to Tyrocidines (Source: ref. 2. Reproduced by kind permission of Kyoritsu Shuppan Co.)

Cyclic decapeptide	Minimum inhibitory concentration (μg/m*l*)		Ref.
	B. subtilis	*S. aureus*	
Gramicidin S	2	2	165
Tyrocidine A (TA)	5	10	164, 165
TB	20	50	165
TC	2	2	165
TE ([L-Phe^{10}]-TA)	5	10	161, 162
[Gly^6]-TA	50	>50	164
[Gly^7]-TA	20	50	164

of natural homologs of tyrocidines. A typical homolog of tyrocidines, tyrocidine A or B, exhibited somewhat weaker activity than gramicidin S.[164,165] This weak activity may be attributed to the basic character of tyrocidines, as in the case of weakly active [L-Orn(Ac)^2]-gramicidin S (No. 2 in Table 4.7): the tyrocidine molecule contains only one basic amino acid residue, ornithine. Analogs in which L- or D-phenylalanine at position 6 or 7 in the tyrocidine A molecule is replaced by glycine exhibit rather weaker activity than the parent tyrocidine A, and some differences in the activities of these tyrocidine A analogs have been observed.[164] The activity of [Gly^7]-tyrocidine A is slightly greater than that of [Gly^6]-tyrocidine A. The differences of specific activity between these analogs may be explained in terms of their hydrophobicity. Droplet countercurrent chromatography of these tyrocidine A analogs showed that the hydrophobicity of the more active [Gly^7]-tyrocidine A is greater than that of [Gly^6]-tyrocidine A and is close to that of tyrocidine A (Fig. 4.2).[164] Precise studies on the mode of antibacterial action and on the conformations of tyrocidines and their analogs will contribute to a better understanding of the structure-activity relationships not only of tyrocidines but also of gramicidin S, because tyrocidines and gramicidin S are intimately related "hybrid" analogs.

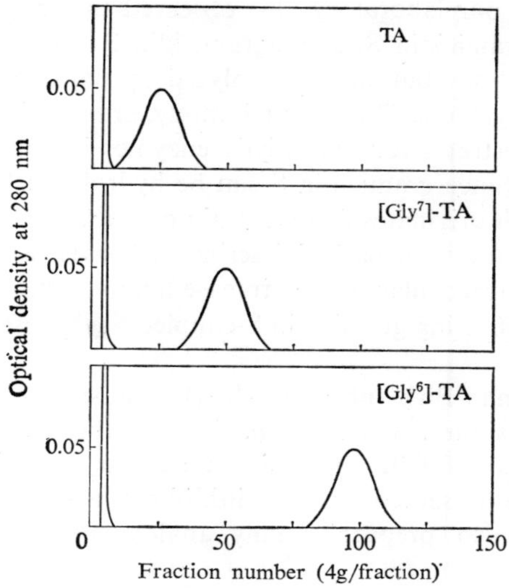

Fig. 4.2 Droplet countercurrent chromatography of tyrocidine A and its analogs. TA, tyrocidine A.

4.5 Utilization of Gramicidin S and Its Analogs in Other Fields

In early work on the structural elucidation of gramicidin S, attempts were made to use usual proteolytic enzymes to rupture the peptide bonds of gramicidin S. However, the antibiotic was resistant to enzymatic hydrolysis. In 1966, Otani *et al.*[220] reported that gramicidin S could be hydrolyzed by nagarse at the peptide linkage between L-valine and L-ornithine, and this hydrolysis afforded two moles of the open chain pentapeptide (**4.25**), as shown in the following scheme. It is interesting to note that gramicidin S, a microbial

$$\text{gramicidin S} \xrightarrow{\text{nagarse}} \text{H–L-Orn–L-Leu–D-Phe–L-Pro–L-Val–OH}$$

(4.25)

metabolite, is resistant to enzymes of animal origin, but is cleaved by the microbial enzyme nagarse, *viz.* subtilisin BPN′. Recently Takahashi *et al.*[221] synthesized the pentapeptide (**4.25**) and compared it with the hydrolysate of gramicidin S by nagarse. The physical properties of the synthetic product (**4.25**) were shown to be in good agreement with those of the peptide obtained from the proteolytic digest.

Ando *et al.* in our laboratory also observed that several gramicidin S analogs, [Gly$^{5,5'}$]-gramicidin S, sesquigramicidin S and digramicidin S, resisted peptic hydrolysis, but were hydrolyzed by nagarse to afford corresponding cleaved fragments. Thus, though rarely, gramicidin S and its analogs can be used as substrates for proteolytic enzymes in spite of their unique primary structure. Since gramicidin S can be hydrolyzed by the serine protease nagarse, which originates from a strain of *Bacillus subtilis*, it may bind to or have affinity for such bacterial serine proteases. Recently Izotova *et al.*[222] isolated an intracellular serine protease homologous with nagarse from *Bacillus subtilis* A-50 using gramicidin S-coupled Sepharose 4B as an affinity support.

Soybean Bowman-Birk inhibitor (BBI) is a small protein consisting of 71 amino acid residues. Nishino *et al.*[151,223] synthesized a cyclic nonapeptide fragment (**4.26**) of BBI containing an antitryptic site by the solid-phase method and observed distinct inhibitory activity towards trypsin. Terada *et al.*[224] also prepared many analogs of **4.26** by the solution

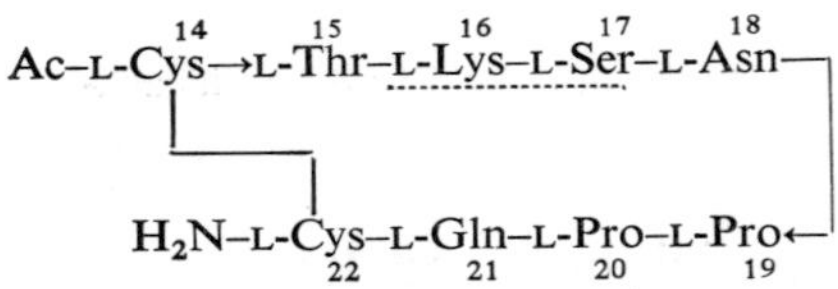

(4.26)

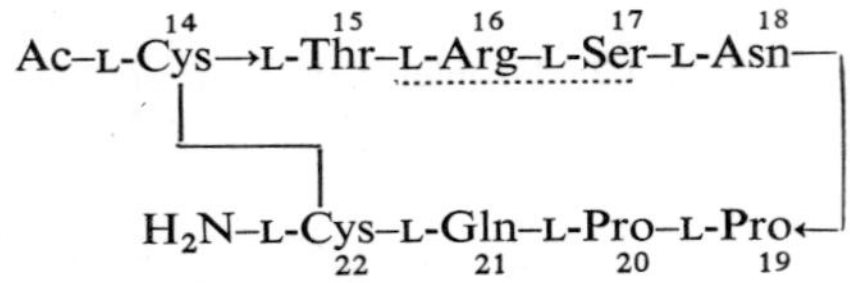

(4.27)

method. One (**4.27**) of the analogs which showed antitryptic activity as strong as that of **4.26** is shown. On the other hand, [L-Arg$^{2,2'}$]-gramicidin S (**4.28**), which has similar, possibly susceptible bonds for tryptic cleavage

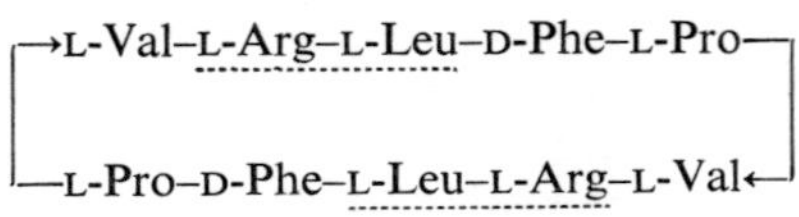

(4.28)

and a similar cyclic structure compared with the fragment (**4.26** or **4.27**) of BBI had no antitryptic activity.[151] The above results suggest that the susceptible bond and cyclic structure of the nonapeptide (**4.26**) and/or the antitryptic site of BBI play a specific role in the inhibitory activity.

It is interesting to note that di-dinitrophenyl-gramicidin S could be used as a dinitrophenyl-antigen which was capable of stimulating and maintaining the production of a homogeneous antibody in the adult animal.[225]

CHAPTER **5**

Conformations of Gramicidin S and Tyrocidines

Gramicidin S has long been of particular interest in conformational studies of peptides. As the distinct difference in biological activity between gramicidin S and its linear decapeptide clearly indicated the importance of the cyclic structure, many investigators attempted to elucidate the three-dimensional structure of the peptide, with or without the benefit of experimental evidence. It is surprising that, in spite of the relative simplicity of the molecule, various different conformations were proposed for gramicidin S and disputed inconclusively for many years. The results of very recent studies, however, have converged on one of the earliest models, a β-pleated sheet structure with C_2 symmetry, which was proposed by Hodgkin and Oughton on the basis of X-ray studies and by Schwyzer based on chemical synthesis. Thus, gramicidin S and tyrocidines can now be used as model peptides to verify the usefulness and reliability of newly developed methods for investigating the conformations of biopolymers.

We will review the results of conformational studies on gramicidin S by X-ray, ORD and CD, NMR, IR, and theoretical studies, after a brief comment on the description of peptide conformations. Presentation of the tyrocidine conformation will be followed by discussions of the remaining problems concerning the conformation of gramicidin S and its analogs. Biological activity will be referred to frequently, because the structure-activity relationship is our main concern. The prefix L is often omitted for simplicity.

5.1 Description of Peptide Conformations in Terms of Torsion Angles

5.1.1 DEFINITION OF TORSION ANGLES

Two notation systmes have been used in the literature. The older one is the notation of Edsall *et al.*[226] and the new one is according to the IUPAC-IUB Tentative Rules.[227] We will use the latter notation. In this system, torsion angles ϕ and ψ are given as angles C′–N–C_α–C′ and N–C_α–C′–N, respectively, as shown in Fig. 5.1. The value 0° is given to ϕ or ψ in the eclipsed conformation, and clockwise rotation is considered positive. Experimentally,

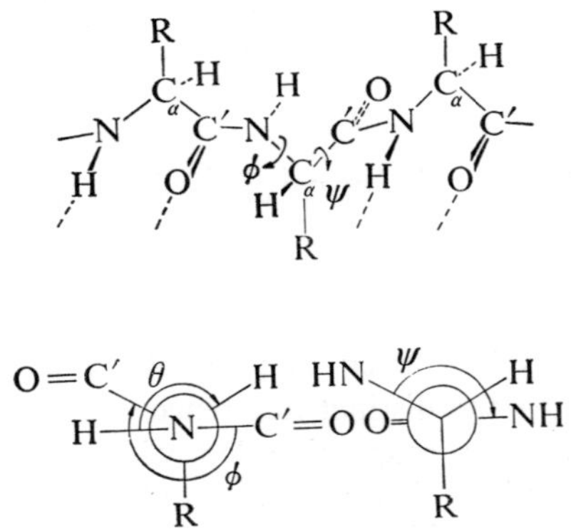

Fig. 5.1 Torsion angles ϕ, ψ, and θ. The line drawing shows a part of the β-sheet structure.

a torsion angle θ, H–N–C_α–H, can be obtained directly from the $^3J_{NHC\alpha H}$ coupling constant of an amide signal in the ^{1}H-NMR spectrum by use of the Karplus-Bystrov equation, which will be described in Section 5.4.1, and θ is given by Eq. (1).

$$\phi = \theta + 60° \qquad (1)$$

On the other hand, another torsion angle ψ can be assessed only by model building. It may be possible to use ^{15}N-NMR for direct estimation of ψ values in future.

5.1.2 DESCRIPTION OF β-TURN CONFORMATIONS

There are a large number of cases in proteins and polypeptides in which the peptide chain reverses its direction. This turning segment has been named a β-turn, β-loop, β-bend, hairpin bend, U-bend, or reverse turn. In the cross β-structure in fibrous proteins, such a turn can occur in the region where a chain folds back on itself. The β-turn is composed of four amino acid residues, the carbonyl group of the first amino acid residue and the amino hydrogen in the fourth residue frequently forming an intramolecular hydrogen bond.

Venkatachalam calculated the allowed conformations of various turns and described the geometries of 15 typical turns.[218)] Among the 15 turns, three broad regions, numbered as No. 2, No. 14, and No. 3, where an intramolecular hydrogen bond exists, are named types I, II, and III.[218)] In addition, mirror images of the allowed turns are also allowed. For example, mirror images of type II, type III, and No. 13 turns are possible and are named type II′, type III′, and No. 13′, respectively. Torsion angles of these turns are shown in Table 5.1. In these typical turns, the second peptide bond is roughly perpendicular to the average turn plane and the carbonyl oxygen is directed downwards in type I, and upwards in type II. Figure 5.2 shows that in type II′ the second peptide bond has a downwards directed carbonyl group because it is a mirror image of type II.

TABLE 5.1 Torsion Angles of Typical β-Turn Conformations

Type	ϕ_2	ψ_2	ϕ_3	ψ_3
I	−60	−30	−90	0
II	−60	120	90	0
II′†	60	−120	−90	0
III	−60	−30	−60	−30
13′†	60	−120	−60	−30

† Type II′ and 13′ turns are the mirror images of type II and 13 turns, respectively.

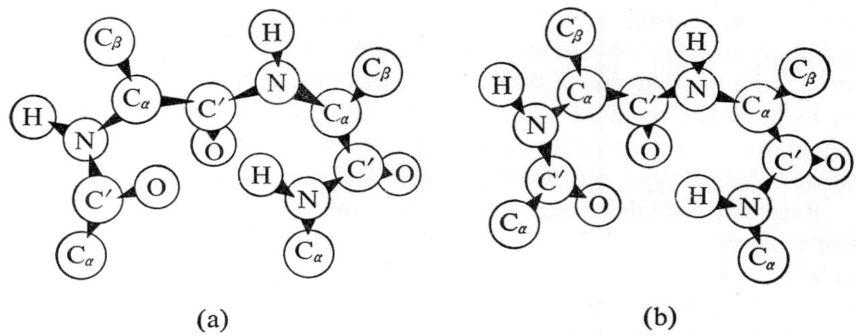

Fig. 5.2 Conformations of typical β-turns. Type II′ (a) and Type III (b).

The tetrapeptide sequence Leu–D-Phe–Pro–Val in gramicidin S forms a β-turn and its precise structure has been determined recently to be very similar to No. 13′ or type II′, as will be shown in Sections 5.3.1, 5.4.1, and 5.5.2.

5.2 X-Ray Crystallography of Gramicidin S and Its Analogs

The first molecular models of gramicidin S based on X-ray analyses of several derivatives[26] were proposed by Hodgkin and Oughton.[229] Although precise conclusions on the size and shape of gramicidin S crystals were not given, they presented several possible models such as α-ribbon, α-helical, β-chain, and mixed α-β models. The results of the X-ray studies suggested that all these models should have a twofold axis of symmetry and, in the crystal, the molecules should be arranged approximately in layers 0.48 nm apart in a direction at right angles to the twofold axis. Among these models, the β-chain model was suggested to be the most probable one. In this model, two chains, each composed of a Val^1–Orn^2–Leu^3–D-Phe^4 peptide unit, can be set antiparallel to one another and linked at the ends by the proline residues to form a decapeptide ring. Four intramolecular hydrogen bonds are formed between the two antiparallel chains. Figure 3.3 in Section 3.1.1 shows this model with slight modifications in the side chains. The implications of this model for the exhibition of biological activity were discussed in Chapter 4.

On the other hand, a mixed α-β model can be constructed from partly α, partly β forms, such as 1) valine to ornithine, α-type (the N–H bonds point in the same direction, and perpendicular to the plane of the ring), 2) ornithine to leucine, β-type, (two N–H bonds point in opposite directions), 3) leucine to phenylalanine, α-type. This model has a true twofold symmetry axis and is strain-free. The third model, the α_{11}-ribbon, is similar to that proposed by Abbott and Ambrose based on IR studies.[230] The original model did not have a strict twofold axis of symmetry. It was revised by Hodgkin and Oughton to have the required symmetry axis,[229] but this modified model was incompatible with the observed strong 0.48 nm reflections. The last model was an α-helical type, but this model also lacked a genuine twofold axis.

Recently, crystals of bis-salicylaldiminato-gramicidin S copper complex were analyzed by Liquori *et al.*[231] Their conclusion accorded with the β-chain model of Hodgkin and Oughton. This result provided further indirect evidence for the existence of β-chain structure in gramicidin S.

Hodgkin and Oughton also proposed a cyclopentapeptide model with one turn of α-helical structure as a possible model of gramicidin S.[229] Therefore, this model should not be considered as a real one of semigramicidin S because this model was proposed on the basis of experiments with gramicidin S. Keeping this in mind, we will discuss it briefly. In this model, all the side chains of L-residues project from one side of the ring and a side chain of D-phenylalanine from the other, such a side-chain orientation being incompatible with the hydrophilic-hydrophobic sided structure mentioned in Section

4.3.1. In this regard, this model may account well for the inactivity of semi-gramicidin S.

Although X-ray crystallography has had great success in analyses of protein structures and small organic compounds of low molecular weight, it is disappointing that X-ray analysis could not give unequivocal results in the conformational analysis of gramicidin S. One reason for this unexpected situation is that gramicidin S crystallizes poorly. The β-chain structure of gramicidin S proposed by Hodgkin and Oughton and independently by Schwyzer based on chemical synthesis studies was finally confirmed by the results of NMR spectroscopy (Section 5.4.1) and energy minimization calculations (Section 5.6.1). As the β-chain structure is a part of the antiparallel β-pleated sheet structure, we will refer to it as β-sheet structure hereafter.

5.3 ORD and CD of Gramicidin S and Its Analogs

As shown in Section 4.3.1, substitution of the constituent amino acids in gramicidin S frequently alters the biological activity, and such an alteration is related to the change in molecular conformation, which can be observed directly by ORD and CD measurements. These optical methods cannot indicate the precise molecular architecture, but they have distinct advantages, such as a simple procedure and sensitivity, and in particular, their high sensitivity to conformational change makes these methods useful tools to detect subtle conformational changes. Figures 5.3. and 5.4. show ORD and CD curves of gramicidin S in ethanol and in 6 M urea in 50% aqueous ethanol.

5.3.1 HISTORICAL REVIEW

The first ORD study of gramicidin S and tyrocidine B was reported by Ruttenberg *et al.*[13)] They observed a distinct Cotton effect in the 200–350 nm region. The dispersion data were used to deduce b_0 values in the Moffitt-Yang[232)] equation, Eq. (2). Here a_0, b_0, and λ_0 are empirical constants, and

$$[m']_\lambda = \frac{a_0\lambda_0^2}{\lambda^2 - \lambda_0^2} + \frac{b_0\lambda_0^4}{(\lambda^2 - \lambda_0^2)^2} \qquad (2)$$

$[m']_\lambda$ is the reduced mean residue rotation at wavelength λ. The b_0 values for gramicidin S, tyrocidine B and linear tyrocidine B were calculated to be—720°, —321°, and —49°, respectively. When these b_0 values and the depths of the troughs at 233 nm were used as measures of helicity, the results (e.g., more than 100% helicity in gramicidin S) were puzzling, because a cyclic decapeptide cannot have a fully helical structure. As the resemblance to helical struc-

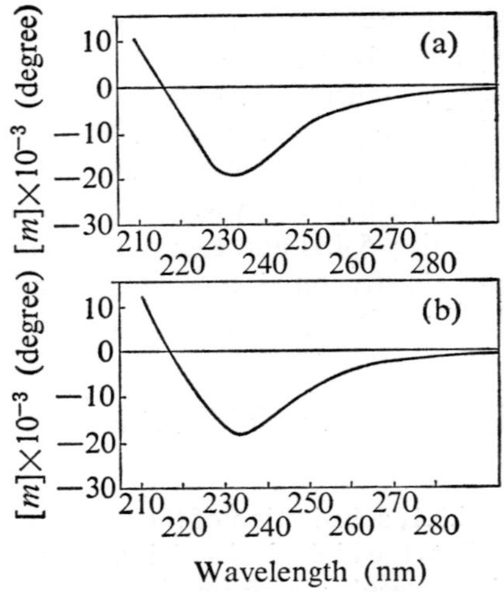

Fig. 5.3 ORD of gramicidin S in ethanol (a) and in 6 M urea in 50% ethanol (b).

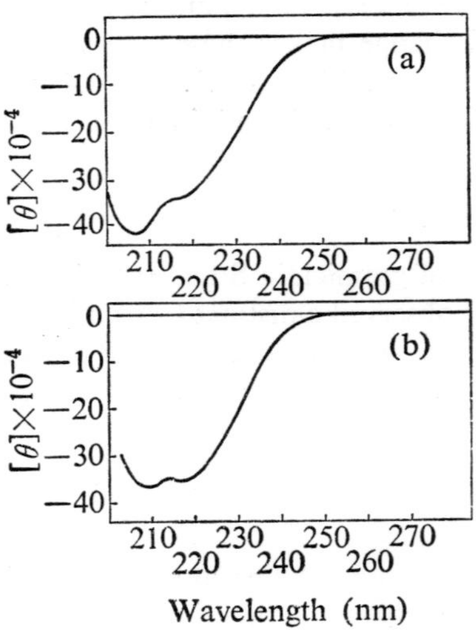

Fig. 5.4 CD of gramicidin S in ethanol (a) and in 6 M urea in 50% ethanol (b).

ture was clear, this contradictory result caused considerable confusion for years.

Quadrifoglio and Urry also studied the ORD and CD of gramicidin S and tyrocidines.[233] The ORD curve of gramicidin S showed a trough at 232–233 nm with a mean residue rotation (uncorrected for the refractive index) of $-18{,}000°$, a shoulder at 210–212 nm, and a peak at 190–191 nm of 60,000° in water. The CD spectrum of gramicidin S had two negative bands at 208 and 217 nm with an ellipticity value on a mean residue basis of -3×10^4 deg·cm^2/decimole. They stated that the complexity of the spectrum and the large ellipticity values of the bands suggested a highly ordered structure. They supported the α-helical structure predicted by Liquori *et al.*,[234] which contained eight helical residues in two separate α-helical turns out of the ten residues in gramicidin S. Urry *et al.* further observed that this CD pattern was essentially unchanged upon hydrogenation.[235] This means that the characteristic trough at 207–217 nm is due to backbone conformation.

Balasubramanian studied the ORD and CD of gramicidin S and [sarcosine$^{5,5'}$]-gramicidin S.[91] He showed that the spectra of these peptides are very similar and that the dispersion profile indeed resembles that of a right-handed α-helix, but closer inspection of the peak positions, the cross-over points, and the magnitudes of the rotations suggested that such a helix is not necessarily present.[236] In particular, the cross-over point of gramicidin S and the analog, at 215 nm, is quite different from that encountered in either a complete α-helix or a complete β-form. The structure cannot be a mixture of α and β forms, or a mixture of random coil with either of these conformations. He predicted that the specific ORD pattern of gramicidin S might be derived from its rigid cyclic structure and he supported the mixed α-β structure proposed by Hodgkin and Oughton.[229]

Laiken *et al.*[166] suggested that the results of NMR studies on gramicidin S[237] ruled out a conformation containing α-helical structure in spite of similarlity in the CD patterns. Laiken *et al.* therefore claimed that care is required in applying ORD or CD to assess helix content in proteins or polypeptides, especially those containing cyclic structures.[166]

In 1974, Woody predicted CD curves for various β-turn conformations on the basis of theoretical calculations.[238] He considered 15 turn conformations proposed by Venkatachalam[218] and conformations suggested by Chandrasekaran *et al.*[217] Chandrasekaran *et al.* suggested that if a dipeptide unit with a D–L sequence formed a β-turn, the most stable dihedral angles ϕ_2, ψ_2, and ϕ_3, ψ_3, of the bend were most probably $+60°$, $-100°$, and $-60°$, $-40°$. Such a turn is very similar to the turn No. 13′ or type II′ defined by Venkatachalam, as shown in Table 5.1. Woody showed in his calculations that a bend No. 13′ should have a 50% probability of giving an α-helix-like CD pattern. Thus, the D-Phe–Pro sequence in gramicidin S has a high probability of exhibiting an α-helix-like CD pattern if its geometry is akin to No. 13′. These torsion angles mentioned above accord with the angles obtained by NMR experiments, as will be described in Section 5.4.1. This calculation showed that the CD curve observed for gramicidin S could be explained by the β-turn at D-Phe–Pro in conformation No. 13′, although the magnitude of the suggested ellipticity was much lower than the experimentally observed values. Old predictions by Balasubramanian and by Craig *et al.* that an α-helix-like ORD pattern of gramicidin S might be derived from specific β-turn segments were thus confirmed, at least qualitatively. Theoretical optical calculations and experimental ORD and CD results supported the conclusion obtained by NMR studies ten years after the first ORD study.

5.3.2 RELATIONSHIP BETWEEN ORD CURVES AND BIOLOGICAL ACTIVITY

The results of our ORD studies on amino acid substituted analogs of gramicidin S suggest that the ORD curves of the analogs can be classified to four types, as shown in Fig. 5.5.[2,239] The relationship between biological activity and the types of ORD curves of these analogs can be summarized as

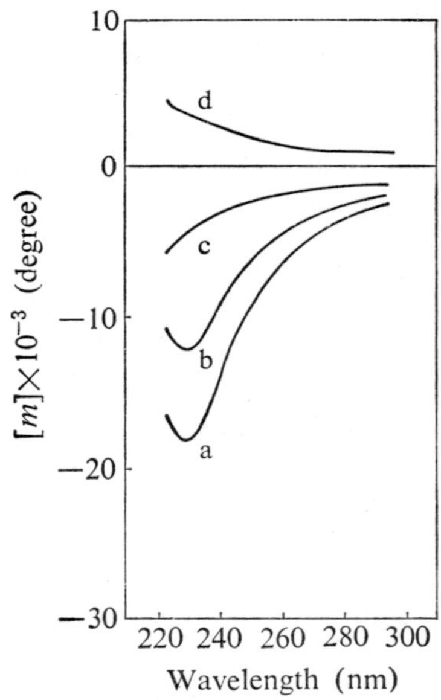

Fig. 5.5 Typical ORD curves of gramicidin S analogs (see the text for details). (Source: ref. 2. Reprodced by kind permission of Kyoritsu Shuppan Co.)

follows. (1) Almost all analogs exhibiting strong biological activity show ORD curves of type a in Fig. 5.5. One exception is [Gly^5]-gramicidin S,[89] showing a type b curve, but no active anlaogs show curves of type c or type d. (2) Analogs with weak activity show usually curves of type b. Curves of type a were observed occasionally, but type c and type d were never observed for active analogs. (3) Inactive or marginally active analogs show type c and type d curves. For example, inactive semigramicidin S shows a curve of type c.[239] However, type a or type b curves also appeared occasionally. In other words, the presence of a trough in the 233 nm region is a necessary condition for an analog to exhibit activity but not sufficient condition for activity.

Our summary is compatible with the data obtained by Sōfuku *et al.*[118] They showed that a strongly active analog [δ-aminovaleric acid$^{4-5}$]-gramicidin S and a marginally active analog [δ-aminovaleric acid$^{5-1'}$]-gramicidin S[94] exhibited troughs with depths of −9,000° and −4,000°, respectively. On the other hand, inactive [δ-aminovaleric acid$^{4-5,4'-5'}$]-gramicidin S showed ORD and CD curves similar to those of random structure.[93] These results indicate that the characteristic conformation of gramicidin S is supported not only by the four intramolecular hydrogen bonds but also by the presence of the D-Phe–Pro peptide bond, forming a stable β-turn.

Additional support for this view came from the synthesis of analogs containing hydroxy acids. As described in Section 4.3.3, Zhuze *et al.*[200] synthesized two analogs, one containing two residues of L-2-hydroxy-3-methylbutanoic acid in place of valine, and the other containing two L-2-hydroxy-3-methylbutanoic acid residues and two L-2-hydroxy-4-methylpentanoic acid residues in place of valine and leucine, respectively. The biological activity of the former was some 5–25% of that of gramicidin S but that of the latter disappeared completely. These changes in the activity paralleled the changes in their CD curves. The former compound showed a shallow trough in the 200–215 nm region but the latter did not.

As shown in Section 5.3.1, the β-turn at the D-Phe–Pro sequence in gramicidin S seems to give rise to a characteristic trough at 233 nm in the ORD curve. If we postulate that this rigid backbone system is a necessary condition for biological activity of gramicidin S, we can easily interpret the presence of curves of type c or type d in Fig. 5.5, observed for inactive analogs, as representing a loss of the stable β-turn in the corresponding sequence. Thus, measurement of ORD or CD curves can be used as a simple determinant to eliminate inactive analogs.

We can conclude that an active analog should have at least one stable β-turn, and that the presence of this structure can be verified by the presence of a deep trough in the 233 nm region in its ORD curve. The fact that inactive *N,N'*-dibenzyloxycarbonyl-gramicidin S shows an intense trough at 233 nm in its ORD curve indicates that the presence of at least one amino group in the side chain of a basic amino acid is an additional prerequisite for biological activity.

5.4 NMR of Gramicidin S and Its Analogs

Figure 5.6 shows the 100 MHz ^{1}H-NMR spectrum of gramicidin S dihydrochloride in dimethylsulfoxide-d_6. Signals of NH_2 protons in the side chains in ornithine are lost in this spectrum because of exchange. Higher resolution has been obtained in 220 MHz spectra in the literature.[237]

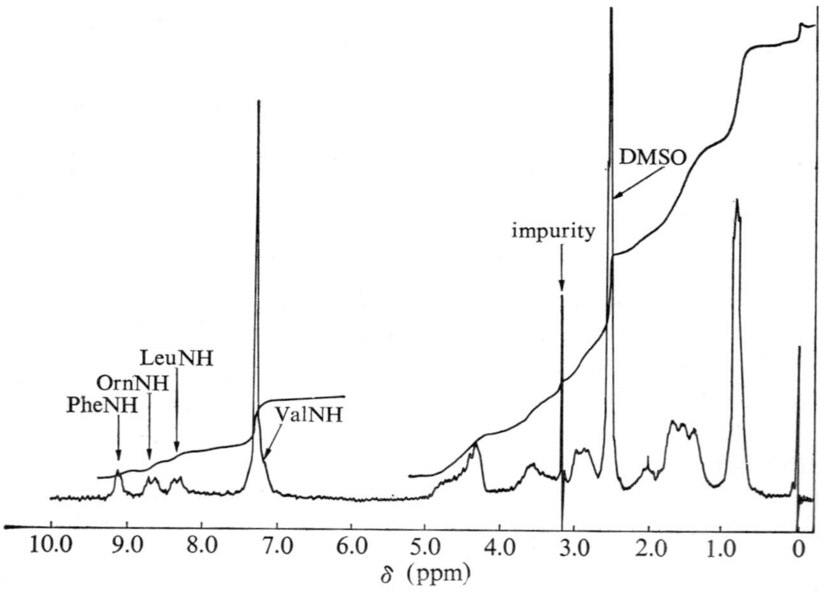

Fig. 5.6 100 MHz NMR spectrum of gramicidin S dihydrochloride in dimethylsulfoxide-d_6.

5.4.1 HISTORICAL REVIEW

Three NMR studies of gramicidin S were reported in 1968. Firstly, Liquori and Conti presented the NMR spectra of gramicidin S and predicted that gramicidin S was in an α-helical conformation,[240] which was compatible with their original conformation calculations.[234] However, the assignment given in this report was later found to be erroneous and their conclusion was withdrawn.[241] Bradbury *et al.* claimed that spectroscopic techniques have given inconclusive results because of the cyclic structure of this peptide and the fact that regular structure, if any is present, can involve only a very small number of residues.[242]

Ten years after the first proposal of β-sheet structure based on chemical synthesis, Schwyzer and Ludescher added new evidence by an NMR study of *N,N'*-diphthaloyl-gramicidin S.[243] Phthaloyl protons of *N,N'*-diphthaloyl-gramicidin S showed splitting and upfield shift in the NMR spectra, suggesting interaction of these protons with phenyl protons in D-Phe residues. They proposed that the only model allowing such an interaction was a β-sheet type structure. However, the conformation of a modified compound might differ from that of the original peptide.

Stern *et al.* measured 60, 100, and 220 MHz NMR of gramicidin S, including assignment of all NH and αCH signals and determination of D-H exchange rates.[237] Anomalously high field shift of the valine NH protons was explained by involving anisotropic shielding by the carbonyl group of D-phenylalanine. Experiments on the rates of D-H exchange indicated intramolecular hydrogen bonding with valine and leucine NH protons. An interpretation of IR data obtained by Abbott and Ambrose is also given in Section 5.5.1. They concluded that among several possible models, including an α-helical one, only the β-sheet structure of Hodgkin and Oughton and of Schwyzer was compatible with the data obtained. They proposed values for all the torsion angles for the first time. The proposed ϕ and ψ values are as follows: Val, Orn, and Leu, $-150°$ and $180°$; D-Phe, $-30°$ and $-50°$; and Pro, $\psi = -50°$. The proposed conformation of the D-Phe–Pro segment is akin to type III of Venkatachalam.[218]

In 1969, Ohnishi and Urry reported temperature-dependence experiments with gramicidin S.[244] Their former conclusion of an α-helix based on ORD experiments was rejected in favor of a β-sheet structure, because the very low temperature dependence of four protons of valine and leucine could be explained only by postulating intramolecular hydrogen bonding in the β-sheet structure. Later in 1972, Pitner and Urry confirmed this conclusion by solvent-dependence experiments.[245] Amide protons of ornithine and D-phenylalanine residues showed dramatic upfield shifts (about 1 ppm) on changing the solvent from methanol to trifluoroethanol. Values for protons of valine and leucine residues shifted only slightly.

Bystrov *et al.* derived a Karplus-like equation, Eq. (3), and elucidated the relationship between coupling constant $^3J_{NHC\alpha H}$ and torsion angle θ.[228]

$$^3J_{NHC\alpha H} = 9.4\cos^2\theta - 1.1\cos\theta + 0.4 \tag{3}$$

This relationship was applied to conformational studies of gramicidin S and diacetyl-gramicidin S by Ovchinnikov *et al.*[246]

Their new approach was a joint use of several independent physicochemical methods such as ORD, IR, and NMR. ORD curves of gramicidin S and diacetyl-gramicidin S showed similar and very small solvent dependence between polar and non-polar solvents, indicating close similarity and comparable rigidity of the two amide chromophore systems. Quantitative IR measurements indicated six hydrogen-bonded amide groups in diacetyl-gramicidin S. In addition, an NMR study on the D-H exchange rates of NH protons revealed four strongly hydrogen-bonded protons of valine and leucine, and two weakly hydrogen-bonded ornithine δ-protons, which interacted with the carbonyl groups of the acetyl groups. Use of the Karplus-Bystrov equation, Eq. (3), for the measured $J_{NHC\alpha H}$ coupling constants indicated *trans* orientations of the NH–C_αH protons for valine, ornithine, and leucine residues, and *gauche* orientation for D-phenylalanine.

On the basis of this derivation and corresponding model building, all the torsion angles of the five amino acid residues were determined as follows: ϕ and ψ values for Val, $-120°$ and $120°$; Orn, $-110°$ and $110°$; Leu, $-120°$ and $110°$; D-Phe, $55°$ and $-110°$; Pro, $-60°$ and $-40°$. The D-Phe–Pro segment is in the type II′ conformation of Venkatachalam. The torsion angles given by Ovchinnikov *et al.* differ from those given by Stern *et al.*, because the values given by the latter group were predicted without consideration of the Karplus-Bystrov relaionship. In 1970, Gibbons *et al.* also applied the Karplus-Bystrov equation in an NMR study, and developed a method to investigate peptide conformations by the combined use of NMR and conformational energy calculations.[247] This combined method has begun a new era in the study of conformations of biologically active peptides.

5.4.2 NMR OF GRAMICIDIN S ANALOGS CONTAINING *N*-METHYLLEUCINE

Analogs containing *N*-methylleucine, such as [L-MeLeu^3]-gramicidin S and [L-MeLeu$^{3,3'}$]-gramicidin S, exhibit strong antimicrobial activity and have an intense trough in the 233 nm region in the ORD spectra.[121,122] These features are the same as those of gramicidin S. As an *N*-methylleucine residue has no amino hydrogen, these peptides cannot be in a β-sheet structure with four intramolecular hydrogen bonds. We considered several models, for example, β-structure containing only two hydrogen bonds of valine protons. However, model building revealed that the bulky *N*-methyl groups cannot be incorporated in the β-structure.

In 1975, Kumar *et al.* proposed conformations for the two *N*-methyl analogs on the basis of NMR studies.[248,249] A model presented by Kumar *et al.* for [L-MeLeu$^{3,3'}$]-gramicidin S in methanol contained two single turns of 3_{10} helices with the helical axis perpendicular to the C_2 symmetry axis, consisting of MeLeu–D-Phe–Pro–Val–Orn residues and amide protons of valine and carbonyl groups in *N*-methylleucine in weak, and protons of ornithine and carbonyl groups in D-phenylalanine in strong intramolecular hydrogen bonds. In this conformation, the directions of the side chains in aliphatic amino acid residues were completely different from those of the original gramicidin S, as shown in Fig. 5.7.

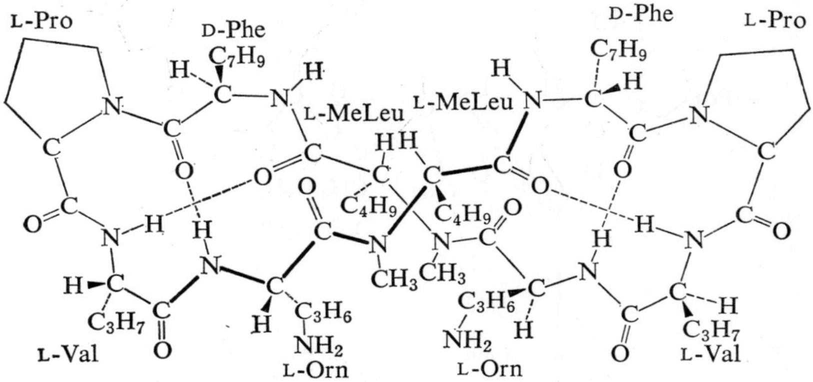

Fig. 5.7 Possible conformation of [L-MeLeu[3,3′]]-gramicidin S in methanol. (Source: ref. 249. Reproduced by kind permission of The American Chemical Society)

As we have seen in the previous sections, almost all of the results in conformational studies are in accord with the β-sheet structure shown in Fig. 3.3. A characteristic feature of this model is a "sidedness", in which four aliphatic side chains are directed to one side of the average ring plane and the charged side chains of ornithine are directed to the other side. Thus the overall orientation of the side chains resembles that of a cationic detergent. Such sidedness was lost in Kumar's models.

A new model with *cis* orientation at the Orn–MeLeu peptide bond was proposed by Patel and Tonelli based on model building and energy minimization calculations.[250)] Usually, *cis* orientation in a peptide bond is unfavorable other than for a proline bond. When *N*-methylleucine replaces leucine, the energistic instability of *cis* orientation must be minimized because of the *N*-substitution. Proposed dihedral angles in this conformation were compatible with the observed NMR data. An additional feature of this model was the eclipse orientation of NH–C_αH hydrogens in ornithine. A new intramolecular hydrogen bond system, involving one carbonyl group in *N*-methylleucine interacting with two amide hydrogens in valine and ornithine, was also postulated. As shown in Fig. 5.8, the orientations of the side chains in this conformation differed from those of gramicidin S and the sidedness is lost.

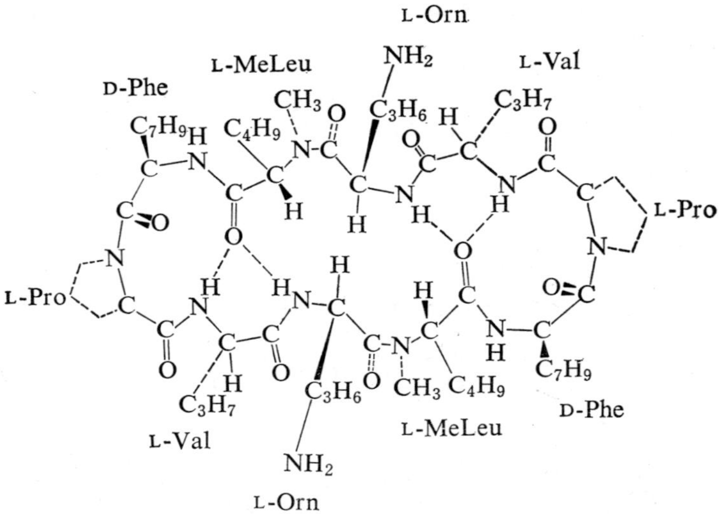

Fig. 5.8 Possible conformation of [L-MeLeu$^{3,3'}$]-gramicidin S in methanol. (Source: ref. 250. Reproduced by kind permission of John Wiley & Sons.)

A third conformation for [L-MeLeu^3]- and [L-MeLeu$^{3,3'}$]-gramicidin S in methanol was proposed by Kato and Izumiya.[219,251] The characteristics of this conformation are as follows: 1) all *trans* peptide backbone system, 2) eclipse orientation for NH–C_αH protons in the ornithine residue, 3) helical or one-to-two hydrogen bond system, 4) conservation of the sidedness, that is, all aliphatic side chains on one side and a cationic charged group in ornithine on the other side of the average ring plane. As shown in Fig. 5.9, the *N*-methyl group in *N*-methylleucine is directed to the hydrophobic side and enhances the hydrophobicity of the plane; this explains the enhanced activity of these *N*-methylated analogs.[219] In all these three conformations, the D-Phe–Pro segment forms β-turns of type II′ or No. 13′, so that all the predicted conformations are compatible with observed ORD and CD data.

Unlike [L-MeLeu^3]- and [L-MeLeu$^{3,3'}$]-gramicidin S, analogs containing *N*-methylvaline exhibited only weak biological activity.[123] The activity of [L-MeVal^1]-gramicidin S was some 1/5–1/10 of that of gramicidin S, and that of [L-MeVal$^{1,1'}$]-gramicidin S was one-half of that of the monomethyl analog. We should point out that hydrogens of the valine residues take part in intramolecular hydrogen bonds in all the predicted conformations. Therefore, loss of the amide hydrogen of valine makes the formation of a stable β-turn difficult. This view was confirmed by the observation that the characteristic trough at 233 nm in the ORD changed to very shallow trough for [L-MeVal^1]-

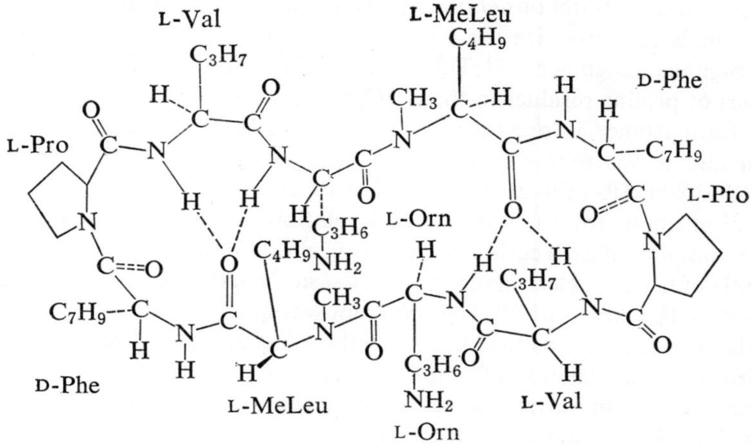

Fig. 5.9 Possible conformation of [L-MeLeu[3,3′]]-gramicidin S in methanol compatible with the "sidedness" hypothesis.

gramicidin S or to a negligible one for [L-MeVal[1,1′]]-gramicidin S. All these features support the view that the sided structure stabilized by rigid β-turns is a prerequisite for antimicrobial activity. As described in Section 5.3.1, these results are compatible with those obtained by Zhuze *et al.*[200)] and Sōfuku *et al.*[29,93,118)]

5.4.3 RECENT NMR STUDIES OF GRAMICIDIN S

Since the conformation of gramicidin S was elucidated, the peptide has been used as a model to demonstrate the usefulness of newly developed experimental procedures. A new technique of intramolecular double resonance (indor) spectroscopy was introduced by Gibbons *et al.* to elucidate secondary (backbone) and tertiary (side chain) conformations.[252)] The major advantage of this method is a great simplification of the entire spectrum, making assignment and interpretation of signals extremely easy.

Nuclear Overhauser effect (NOE) was used by Rae *et al.* to estimate the distances between protons.[253)] Because of slow tumbling of the gramicidin S molecule, negative NOEs were observed. The NOEs between amide protons of ornithine, leucine, and D-phenylalanine and α-protons of valine, ornithine, and leucine, respectively, confirmed the β-sheet structure of gramicidin S. Gibbons *et al.*[254)] detected further NOE between the α-protons of D-phenylalanine and proline and suggested that this observation is consistent with a type II′ β-turn.

The first ^{13}C-NMR measurement and a complete assignment of signals were presented by Gibbons *et al.*[255] Sogn *et al.* used specifically ^{13}C-enriched amino acids to emphasize the corresponding signals, and thus obtained an unambiguous assignment.[256] The chemical shift difference between β and γ carbons of proline residues in these ^{13}C-NMR spectra revealed that only the *trans* form isomer is present.[257] The all-*trans* backbone conformation of gramicidin S was thus established by Bovey.[257] Allerhand and Komorski studied ^{13}C-NMR spin-lattice relaxation of gramicidin S and obtained information about the rates of internal rotation of the side chains.[258] The spatial mobility of each carbon atom was clarified for the first time. As another example of heteronuclear magnetic resonance, ^{15}N-NMR of gramicidin S was studied by Hawkes *et al.*[259] The spectrum was unusually simple, and six ^{15}N signals obtained were assigned. This method was applied by Khaled *et al.* to confirm the conformation of gramicidin S.[260] They assigned ^{15}N resonances of gramicidin S in various solvents by utilizing the secondary structure of gramicidin S and comparing the ^{15}N-spectrum of the peptide with that of [D-MePhe$^{4,4'}$]-gramicidin S.

5.5 IR and Other Experimental Studies of Gramicidin S and Its Analogs

Figure 5.10 shows the IR spectrum of gramicidin S in a KBr disk. It exhibits a strong amide I signal at 1,650 cm^{-1}, and amide II at 1,540 cm^{-1}, but no signal or shoulder at 1,690 cm^{-1}, indicating the presence of all-*trans* amide bonds and the apparent absence of antiparallel β-sheet structure.

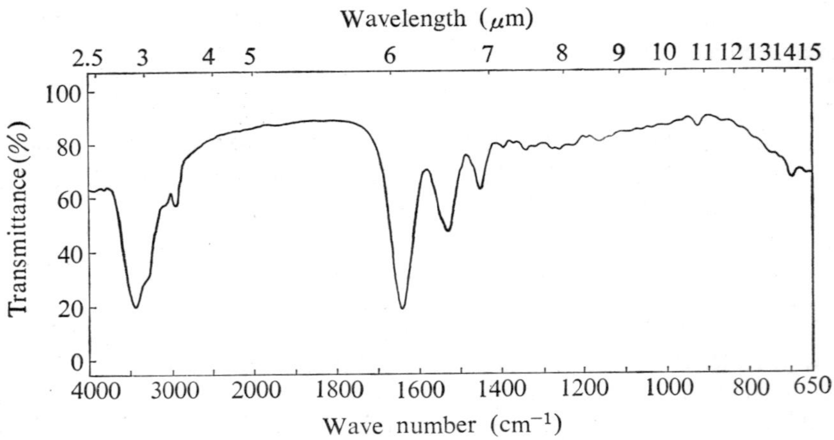

Fig. 5.10 Infrared absorption spectrum of gramicidin S dihydrochloride in a KBr disk.

5.5.1 IR OF GRAMICIDIN S

Crystalline gramicidin S derivatives show unusually intense infrared dichroism in the 4,500–5,000 cm^{-1} region. A folded α-ribbon model was proposed by Abbott and Ambrose with ten units of seven-membered hydrogen bond rings **(5.1)**, and was compatible with this IR result. The main difficulty

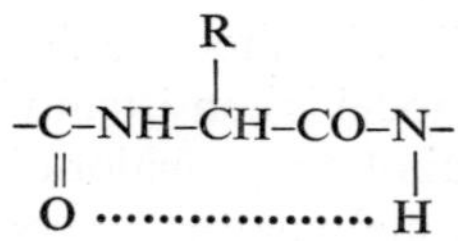

(5.1)

of this model was the absence of a genuine twofold axis of symmetry, this being required by the X-ray crystallography data. This dichroism was later interpreted by Stern *et al.* in terms of a steric factor; that is, at least six peptide bonds are in the plane perpendicular to the C_2 axis, while two phenylalanyl peptide bonds are almost parallel to this axis.

A new IR study appeared 20 years later. Gramicidin S was incorporated in a polyoxyethylene film, which was then uniaxially stretched.[261] The infrared characteristics of the stretched film allow measurements of the dichroic spectra of NH amide stretching between 3,500 and 3,000 cm^{-1} and of amide II bands between 1,700 and 1,500 cm^{-1}. The results of this experiment indicated a weak perpendicularly polarized shoulder at 1,680 cm^{-1} and a strong peak with parallel polarization at 1,640 cm^{-1}. This pattern indicated the presence of cross-β structure. The cross-β dichroic spectrum obtained could arise from association of the β-sheet conformations into ribbon-like aggregates. The results of D-H exchange show that most of the amide groups of gramicidin S are shielded from the solvent by the participation of the secondary amide groups in strong hydrogen bonding interactions. Thus, they proposed that a cross-β aggregate of gramicidin S was formed in oriented polyoxyethylene.

5.5.2 OTHER EXPERIMENTAL METHODS

Countercurrent dialysis was developed by Craig.[262] A thin film of cellophane with adjusted pore size was used to separate the solutes. Gramicidin S assumed not a compact but an open structure, this seeming to support a β-sheet structure rather than a helical or α,β-mixed structure. Laiken *et al.* determined the rates of T-H exchange by means of this dialysis method.[263] They found four rapidly exchanging hydrogens and two groups of two slow hydrogens in the base-catalyzed pH region. These results are entirely consistent with the NMR data of Stern *et al.* and provide additional evidence for a

β-sheet structure. These T-H exchange data measured in aqueous solution were in quite good agreement with the data obtained from NMR spectra in methanol or dimethylsulfoxide solution. This suggests that the conformations in these solvents are similar, if not identical, and that the backbone of the molecule is quite rigid. These data support a β-type structure, but cannot give such clear-cut evidence of precise structure as the NMR data..

Warner proposed a fused hexagon model for gramicidin S. This model was obtained simply by molecular model building without any experimental support.[264] Although this model is consistent with the X-ray results of Hodgkin and Oughton and is strain-free, its major disadvantage is the fact that all the peptide bonds are in an unfavorable *cis* conformation.

5.6 Theoretical Studies of Gramicidin S Conformation

In order to identify the conformation of a biologically active polypeptide with the global minimum of potential energy, one has to search tremendous numbers of conformations with similar energy. Furthermore, it is very difficult to evaluate precisely the interaction of the peptide with the solvent. Therefore, results obtained by calculation should be considered as qualitative. There are two methods of theoretical conformational analysis, semiempirical and quantum-mechanical. Both methods are cumbersome and require much time for computation. However, these procedures may have considerable potential in the future. Ovchinnikov and Ivanov described the present status of theoretical studies on the conformational analysis of cyclic peptides as follows: "whatever the difficulties and shortcomings of the computational methods, a rational combination of the theoretical and experimental techniques will, on agreement of the two approaches, give greater confidence in the results obtained."[201]

In 1966, Liquori *et al.* proposed a molecular model of gramicidin S based on theoretical energy calculation.[234] This model was built up from five peptide units which were fixed in helical conformational states. The resulting model consisted of two antiparallel single turns of right-handed α-helices bridged by proline and D-phenylalanine.[239] Each helix contained one intramolecular hydrogen bond between NH of D-phenylalanine and the carbonyl group of valine. There is a C_2 symmetry axis perpendicular to the two axes of the helices. After this model had been withdrawn, a revised model was proposed in 1971 using similar peptide units and renewed energy functinos.[265] The results of NMR experiments such as $J_{NHC\alpha H}$ coupling constants were used to limit the numbers of possible conformations. The conformation obtained in this manner was a β-pleated sheet structure of the Hodgkin-Oughton-Schwyzer form with four hydrogen bonds. Proposed ϕ and ψ values for each torsion angle were as follows: Val, −82° and 137°; Orn, −132° and 142°; Leu, −143° and 82°; D-Phe, 58° and −116°; Pro, −68° and −31°.[265]

Semiempirical studies on conformation calculations of gramicidin S were mainly carried out by Scheraga's group. Four papers were published from 1967 to 1975 and in every case considerable progress and revision of the conclusions were apparent. The first model was a β-structure-like model.[266] This model has only two intramolecular hydrogen bonds, and was selected from 282 conformers. In the second study, the first model and the α-helical model of Liquori *et al.* were used as the starting models and minimization calculations were carried out.[267] The models obtained were a β-like structure with two intramolecular hydrogen bonds and an α-helical structure without hydrogen bonds. In the third study, the β-structure of Hodgkin-Oughton-Schwyzer type was taken as the starting point for the minimization procedure.[268] The resulting structure was a β-like one without hydrogen bonds.

In the final study, Digert *et al.* proposed a β-sheet conformation with six hydrogen bonds, the side chains of the ornithine residues extending near to the D-Phe residues, and the δ-NH group of ornithine interacting with the carbonyl group of D-phenylalanine.[269] This model had the lowest energy so far calculated, even when the stabilizing effect of the two additional hydrogen bonds between δ-NH in ornithine and the carbonyl groups in D-phenylalanine was neglected. The predicted conformation was shown to be compatible with currently available experimental data. This model was further compared with previously proposed models. The model proposed by Stern *et al.*[237] contained two pleated sheet segments joined by a type III turn. On the other hand, the model proposed by Ovchinnikov *et al.*[246] included a type II′ turn. These turns are shown in Fig. 5.2 (Section 5.1). A revised model of De Santis and Liquori[265] also contained a type II′ turn. The model by Digert *et al.* also had a type II′ β-turn, with torsion angles ϕ and ψ as shown below: Val, −90° and 100°; Orn, −127° and 125°; Leu, −156° and 117°; D-Phe, 60° and −139°; Pro, −75° and −18°.

Pysh developed a new method named optical calculation for screening low-energy conformations obtained from conformation calculations.[270] Several conformations proposed for gramicidin S were screened. Only one type led to a significantly large negative b_0 value. This geometry is the β-sheet type structure proposed by Momany *et al.*[268] On the other hand, two helical models yielded very small positive b_0 values. These results seem to be compatible with those of Woody as described in Section 5.3.1. Based on these results, Pysh concluded that parameters derived from polymer spectra, and especially the b_0 scale of helix content should probably not be used to describe small peptides, particularly cyclic ones. This conclusion is now quite familiar.

5.7 Conformation of Tyrocidines

In the first CD study of tyrocidines and hydrogenated tyrocidines, Craig *et al.*[166] suggested that the conformation of tyrocidine B is similar to that of gramicidin S. The fact that the depth of the trough at 233 nm was one-half of that of gramicidin S was consistent with their prediction that the D-Phe–Pro sequence of gramicidin S and tyrocidines was the origin of the characteristic trough, as shown in Section 5.3.1.

Urry pointed out that in solutions containing a large amount of dioxane or acetonitrile, both natural and hydrogenated tyrocidine B exhibited β-structure-like CD patterns, suggesting a possible conformational change in the chromophores.[235] A tetrapeptide sequence, Asn–Gln–Tyr–Val, in tyrocidine B may form a β-turn, but the turn formed from the all-L sequence might be less stable than that of Leu–D-Phe–Pro–Val in gramicidin S. ^{1}H-NMR of tyrocidine B in D_2O, again studied by Craig's group,[271] indicated a strong tendency for aggregation. In an 8% aqueous solution, no NMR signals were observed. On lowering the concentration, raising the temperature, or on addition of methanol, a spectrum could be detected and the resolution improved. All these results indicate aggregation of tyrocidine B in aqueous solution.

Two new techniques, correlation NMR and double resonance difference

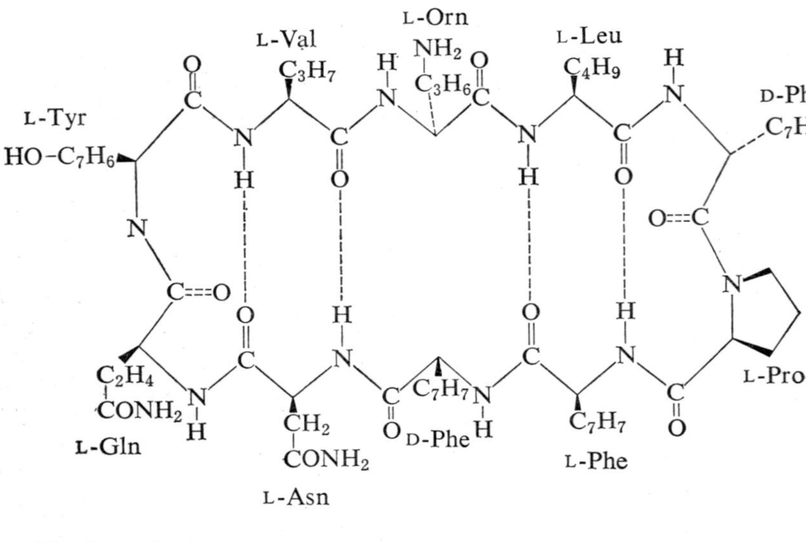

Fig. 5.11 Conformation of tyrocidine A.
(Source: ref. 272. Reproduced by kind permission of The American Chemical Society)

(drd) NMR, were applied to tyrocidines. Correlation NMR provided spectra of good resolution and higher signal-to-noise ratio than could be obtained by the conventional continuous-wave method.[272] Signals of α-protons of ornithine and D-phenylalanine are anomalously downfield in comparison with the signals of the other eigtht α-protons. Such anomalous shifts might result from the positioning of an aromatic ring over these two α-protons. Use of drd spectra in the correlation mode simplifies the whole spectrum of tyrocidine A to the individual spectra of constituent amino acid residues. Complete assignment of the complicated signals of tyrocidine A was achieved for the first time by this method. The conformation of tyrocidine A derived from this study is shown in Fig. 5.11. Six residues of the backbone (Val^1–Orn^2–Leu^3 and Phe^6–D-Phe^7–Asn^8) form an antiparallel β-sheet structure with four intramolecular hydrogen bonds, and the antiparallel segments are joined at each of the two ends by β-turns. On one side of the average ring plane, two side chains, those of L-ornithine and D-phenylalanine, are located, and on the other side are the eight other side chains, as shown in Fig. 5.11. Elucidation of the precise geometry, especially the geometry of the other β-turn at Asn^8–Gln^9–Tyr^{10}–Val^1, is expected to be possible.

5.8 Problems Concerning the Conformation of Gramicidin S

The relationship between antimicrobial activity and variation of the primary structure of gramicidin S was discussed in the previous chapter. In this section, we will consider several analogs designed with particular conformational considerations in view.

5.8.1 MOLECULAR ASSEMBLY OF GRAMICIDIN S: ANALOGS CONTAINING AN INTERMOLECULAR DISULFIDE BRIDGE

No information as yet has been obtained on how gramicidin S molecules assemble in the crystalline or solution state. The molecules may be arranged in parallel or face-to-face. The synthesis of analogs arranged obligately in such assemblies with an intermolecular covalent bond may shed light on this question. For example, it is of interest to bind two molecules of gramicidin S by the use of a bifunctional reagent. Ludescher and Schwyzer synthesized [L-cystine$^{2,2'}$]-gramicidin S and investigated the helicity of the intramolecular disulfide bond.[273,274] Such a disulfide bond can be used to bind two molecules intermolecularly.

Sato *et al.* synthesized [L-Cys(MeOBzl)^1]-gramicidin S and showed that this analog exhibited activity similar to that of native gramicidin S.[133] Synthesis of the desired dimerized gramicidin S derivative by cleavage of the protecting group and successive oxidation has recently been carried out.[133] The possible conformation of the product, *S,S'*-bi ([L-hemicystine[1]]-gramicidin S), is shown in Fig. 5.12. Four hydrophilic ornithine residues are outside of the molecule, and all the hydrophobic aliphatic side chains are buried inside. This appears to explain the inactivity of the dimerized products.[133] The synthesis of molecular assemblies of tyrocidines should also be interesting because tyrocidines are known to exhibit a strong tendency to aggregate in aqueous media due to hydrophobic interactions of the side chains.

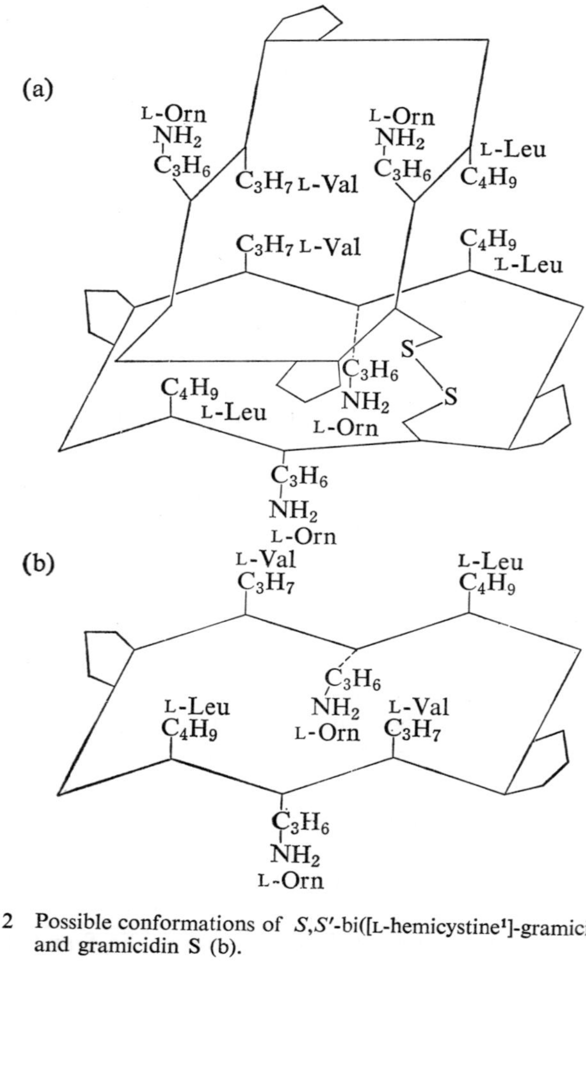

Fig. 5.12 Possible conformations of *S,S'*-bi([L-hemicystine[1]]-gramicidin S) (a) and gramicidin S (b).

5.8.2 TOPOCHEMISTRY OF GRAMICIDIN S

In 1967, Shemyakin *et al.* predicted that the structure of an enantiomer of enniatin, a cyclohexadepsipeptide antibiotic, is very similar to that of the parent enniatin.[275] Synthetic enantio-enniatin exhibited activity identical with that of natural enniatin. They suggested that the effectiveness of a topochemical approach to the structure-activity relationship of biologically active peptides was thus established. In the case of gramicidin S also, it is generally accepted that the essential determinant for biological activity is the three-dimensional structure of the whole molecule. Topochemical study of gramicidin S should verify the reliability of this concept and provide an interesting model to clarify the mechanism of interaction between bioactive peptides and their receptors. As enniatin has high molecular symmetry, the simple enantiomer can exhibit full activity, but in the case of gramicidin S, the situation is more complicated because of its low symmetry.

A. Topochemical Features of Amino Acid Residues

If we assume that the direction of a peptide bond does not influence its biological activity in a biologically active peptide system, in other words, that the activity is related solely to the spatial orientation of the side chains, we would expect the retroenantiomer of the peptide to exhibit the same activity as the parent peptide, because the spatial orientation of the side chains should be the same as that of the parent peptide. (A retroenantiomer is an antipode in which the direction of the peptide bond is reversed.) For example, the spatial orientation of an L-valine residue in the parent peptide is assumed to be the same as that of a D-valine residue in the retroenantio-peptide. The structures of an L-valine residue (**5.2**) and a D-valine residue (**5.3**) can be depicted as follows. Such a spatial similarity can be assumed only for the en-

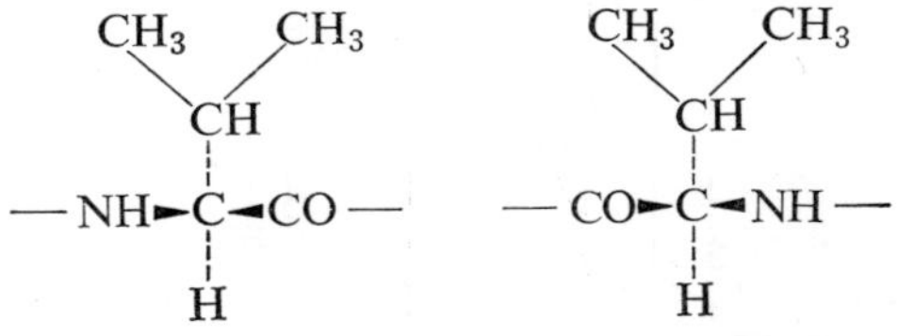

(5.2) **(5.3)**

antiomer, that is an antipode with respect to the α-carbon. If another asymmetric carbon is present, for example, in isoleucine or threonine, the corresponding D-allo-isomer should be used to obtain exactly same "shape" of the molecule. Proline causes another difficulty because of its cyclic structure.

(a)

(b)

(c)

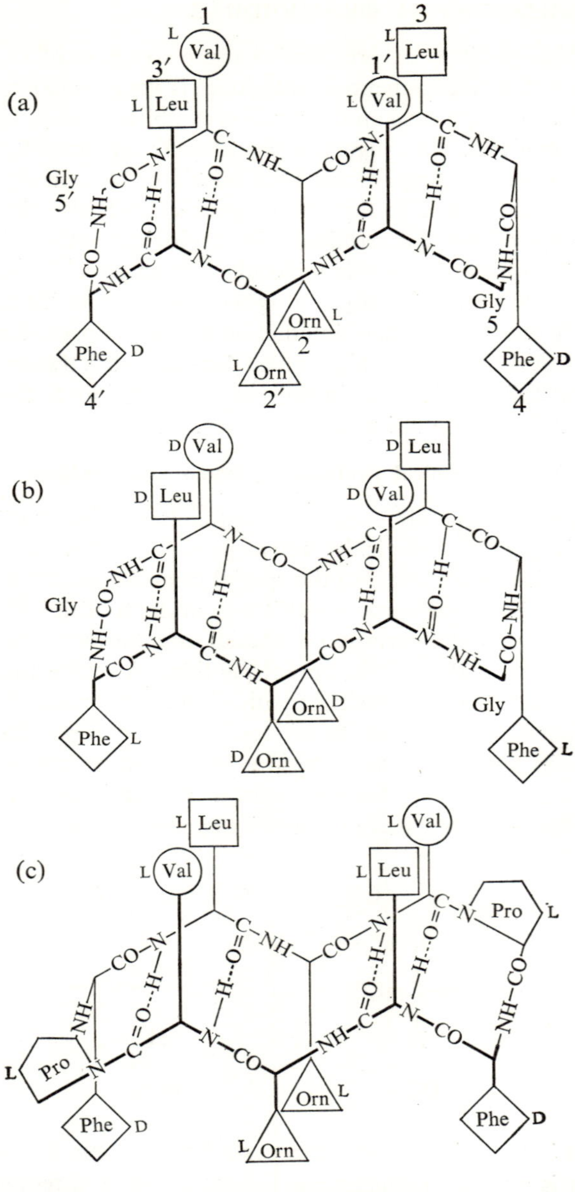

Fig. 5.13 Possible conformations of [Gly$^{5,5'}$]-gramicidin S (a), retroenantio-[Gly$^{5,5'}$]-gramicidin S (b), and retrogramicidin S (c).
(Source: ref. 2. Reproduced by kind permission of Kyoritsu Shuppan Co.)

Reversal of the direction of the peptide bond apparently results in a change in the position and orientation of the pyrrolidine ring in proline.

B. Topochemical Analogs of Gramicidin S

In 1967, Shemyakin *et al.*[146] overcame the difficulty caused from proline by the use of [Gly$^{5,5'}$]-gramicidin S, which had been synthesized by Aoyagi *et al.*[77,78] Synthetic retroenantiomeric [Gly$^{5,5'}$]-gramicidin S was reported to exhibit activity similar to that of the parent peptide.[147]

The presence of an intense trough at 233 nm in the ORD spectrum strongly suggests that the conformation of [Gly$^{5,5'}$]-gramicidin S, shown in Fig. 5.13a, is similar to that of gramicidin S. The conformation of retroenantio-[Gly$^{5,5'}$]-gramicidin S is unknown, but we can arbitrarily suppose it to be as shown in Fig. 5.13b. In this case, the shape of this peptide becomes similar to that of the parent peptide, explaining its activity. Shemyakin *et al.* also stated that retro-[Gly$^{5,5'}$]-gramicidin S showed strong antibacterial activity. As this peptide is a simple retromer, its shape should not be the same as that of the parent peptide. However, it could act as an antibiotic at least functionally. It is reasonable to assume that the relative orientations of the side chains of aliphatic amino acids are similar in these peptides. Waki and Izumiya synthesized retrogramicidin S and showed its activity to be one-tenth of that of natural gramicidin S.[113,114] The supposed conformation of retrogramicidin S (Fig. 5. 13c) differs from that of natural gramicidin S as regards to the position and orientation of the prolyl residue. This difference might affect the correct fit with an assumed receptor and cause lowering of the activity.

5.8.3 ELECTRON PARAMAGNETIC RESONANCE (EPR) OF GRAMICIDIN S AND ITS ANALOGS

Recently Ovchinnikov reported the results of EPR measurements of gramicidin S and its analogs and discussed the relationship between biological activity and the separation of the two δ-amino groups in the ornithine residues.[202] He measured the average distance between two N^1 atoms in the spin-labeled alkyl group **(5.4)** bound to the δ-amino groups of the ornithine

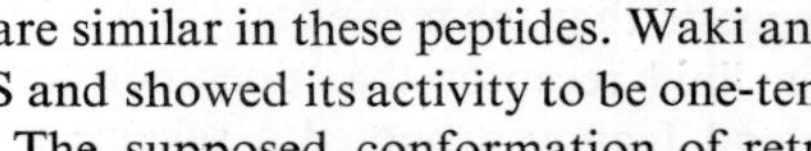

(5.4)

residues in chloroform at −190°. Table 5.2 shows the results, which suggest that the greater the distance between the two aminoalkyl groups is, the less active the analog becomes. An important feature of Table 5.2 is the high

TABLE 5.2 Antimicrobial Activities and Distances between Two δ-Aminoalkyl Groups in Gramicidin S and Its Topochemical Analogs (Source: ref. 2. Reproduced by kind permission of Kyoritsu Shuppan Co. and North-Holland Pub. Co.)

Cyclic decapeptide	Minimum inhibitory concentration (μg/m*l*)		Distance (nm)
	B. subtilis	*S. aureus*	
Gramicidin S (GS)	0.5–1	0.5–1	1.25
EnantioGS	0.5–1	0.5–1	1.25
RetroGS	9–18	60–70	2.2
[Gly$^{5,5'}$]-GS	9–18	18	1.6
Enantio-[Gly$^{5,5'}$]-GS	9–18	18	1.6
Retro-[Gly$^{5,5'}$]-GS	12–18	100	2.2
All-L-GS	9–18	60–75	2.5

activity of enantiogramicidin S. As enantiogramicidin S is the mirror image of gramicidin S, it is hard to understand its full activity if one assumes an enzymatic mechanism (like that of penicillin) for the action of gramicidin S. Instead, this result suggests that the mode of action of gramicidin S involves an interaction with a lipid receptor that does not contain an asymmetric center. Ovchinnikov and Ivanov proposed a scheme for the interaction of gramicidin S with a lecithin membrane.[201)]

It should be noted that these EPR experiments were carried out in non-aqueous media and with protected δ-amino groups, although gramicidin S reacts with sensitive microorganisms in aqueous media and with free δ-amino groups. Although such differences in conditions may constitute defects in this study, this approach seems valuable, since it can give a unique insight into the relationship between microscopic molecular structure and biological activity.

5.8.4 CONFORMATION-ACTIVITY RELATIONSHIP OF GRAMICIDIN S ANALOGS SUBSTITUTED AT THE 4,4′ POSITIONS

Since all-L-gramicidin S is inactive, the biological importance of 4,4′-D-amino acids is obvious. The relationship between activity and amino acid substitution at this 4,4′ turning point has been discussed in Section 4.3.1.E. In order to obtain further insight into this point, Kawai and Nagai synthesized [D-Ala$^{4,4'}$]-, [L-Ala$^{4,4'}$]-, [Gly$^{4,4'}$]-, and [α-aminoisobutyric acid$^{4,4'}$]-gramicidin S and compared their ORD curves.[82,116,276)] ^{1}H-NMR studies of these analogs were carried out by Higashijima *et al.*[277)]

ORD spectra of biologically active D-alanine and glycine analogs show intense troughs of −17,000° and −14,000°, respectively, near 233 nm, whereas inactive L-alanine and α-aminoisobutyric acid analogs show weak minima of $-4{,}000°$ and $-3{,}000°$, respectively. The 270 MHz ^{1}H-NMR spectrum of the D-alanine analog in DMSO-d_6 solution is quite similar to that of gramicidin S. For glycine-gramicidin S, however, some additional resonances are observed in DMSO-d_6 because of the *cis-trans* isomerism about the Gly–Pro peptide bond.[277] The presence of a *cis* peptide bond in gramicidin S analogs was found for the first time. As for the *trans* isomer, the chemical shifts of the main chain protons of the glycine analog are observed between those of the corresponding protons of the D-alanine and L-alanine analogs. This suggests that the conformation of the glycine analog is an ensemble average of the [D-Ala$^{4,4'}$]-gramicidin S-type and [L-Ala$^{4,4'}$]-gramicidin S-type structures, with a mixing ratio of about 3:1, in agreement with the analysis of the ORD spectra: the depth of the trough at 233 nm of the glycine analog is the weighted average of the troughs of the D-alanine and L-alanine analogs.[276]

Kawai and Nagai noted that the methyl group of L-alanine and the *pro*-S methyl group of α-aminoisobutyric acid are on the same side (*pro*-S corresponds to L-side) and the L-side methyl group is absent in D-alanine and glycine.[276] The analog containig L-alanine or α-aminoisobutyric acid (Aib) is less able to take a β-sheet conformation because the L-side methyl group collides with the δ-methylene group of the proline residue when the L-Ala–L-Pro or Aib–L-Pro sequence adopts a type II′ β-turn, as shown in Fig. 3.3. Such steric hindrance prevents these inactive analogs from taking up the β-sheet conformation necessary for antibiotic activity.[276]

CHAPTER 6

Modes of Action of Gramicidin S and Tyrocidines

Gramicidin S and tyrocidines are known to damage the cytoplasmic membrane of sensitive microorganisms and to release nucleotides from the cytoplasm, but studies on the modes of action of these peptides at the molecular level have not progressed significantly for two decades. The reason seems to stem from the fact that the mechanism of action is directly related to the structure and function of the cell membrane of the sensitive bacteria, and is difficult to analyze in terms of molecular interactions. The situation is changing with the recent development of membrane sciences, and studies using refined techniques are in progress. Furthermore, a new biological role of tyrocidines and linear gramicidins has been proposed. These peptides have been targets of study in research on regulation biology in addition to conventional antibiotic studies.

6.1 Antibacterial Action of Polyamino Acids and Linear Peptides Related to Gramicidin S

In 1953, Katchalski *et al.* studied the antibacterial action of lysine and ornithine polymers.[182] In addition to measurements of the minimum inhibitory concentrations of polyamino acids against sensitive bacteria (shown in Table 6.1), they studied the decrease of respiration and electrophoretic mobilities of cells interacting with the polymers, and the adsorption of polyamino acids on the cell surface. Respiration of electronegative organisms such as *Escherichia coli* decreased upon binding with electropositive poly-lysine but the decrease disappeared reversibly on the addition of negatively charged poly-aspartic acid.

In the next study, they synthesized copolymers composed of the amino acids present in gramicidin S and measured their growth-inhibiting activi-

TABLE 6.1 Antimicrobial Activities of Lysine Peptides and Basic Polyamino Acids (Source: ref. 182. Reproduced by kind permission of the Biochemical Society (London))

Peptide	Minimum inhibitory concentration (μg/ml	
	E. coli	*S. aureus*
L-Lys_2	450	
L-Lys_3	150	
L-Lys_4	75	
poly-L-Lys	2.5	0.5–10
poly-DL-Lys	5	3
poly-DL-Orn	10	5
poly-DL-Arg	10	5

ties,[183,184] as shown in Table 4.5. These amino acid copolymers were prepared by polymerization of mixtures of the *N*-carboxy-amino acid anhydrides in the appropriate molar ratios. They observed release of inorganic phosphorus into the media and found that copolymers containing ornithine and leucine react with cells, causing leakage of the cell constituents as a result of an abrupt change in surface permeability. Such copolymers, like gramicidin S, contain hydrophilic charged side chains and hydrophobic alkyl side chains, and were adsorbed at the cell surface, causing changes of the surface structure.

A significant role of hydrophobic amino acid side chains, especially that of leucine, was clearly demonstrated in the case of poly-(Orn, Leu) and poly-(Orn, Leu, Ala). These polymers immediately released the cell contents and caused irreversible damage to the cell. The mode of leakage of inorganic phosphorus caused by poly-(Orn, Leu) or poly-(Orn, Leu, Ala) was similar to that caused by gramicidin S, and the leakage could not be neutralized by the addition of poly-aspartic acid. Furthermore, polymers containing leucine, such as poly-(Orn, Leu), poly-(Orn, Leu, Ala), and poly-(Orn, Leu, Ala_3), showed clear cross-resistance with each other and, to some extent, with gramicidin S. On the other hand, poly-Orn, poly-(Orn, Val), and poly-(Orn, Phe) did not cause immediate leakage of inorganic phosphorus. These polymers did exhibit strong antibacterial activities, but such growth-inhibiting activity was reversibly neutralized by poly-aspartic acid. Poly-Orn, poly-(Orn, Val), and poly-(Orn, Phe) show no cross-resistance with leucine-containing polymers or with gramicidin S.

These results suggest that the mode of action of gramicidin S is similar to that of poly-(Orn, Leu) and other leucine-containing polymers, but differs from that of poly-Orn, poly-(Orn, Val), etc. Microorganisms such as *Escherichia coli* were suggested to combine with leucine-containing polymers such as poly-(Orn, Leu) more strongly than with poly-Orn, poly-(Orn, Val), etc. Here we can see that leucine and valine have a biologically different role in spite of the structural similarity of these two aliphatic amino acids. Such a fine biological distinction had already been observed in the case of analogs of gramicidin S, as described in Sections 4.3.1 and 5.4.2.

It should be noted that the strong growth-inhibiting activities of the amino acid copolymers were observed in synthetic media. When tested in nutrient broth, the inhibitory activities of the copolymers were in most cases 5–10 times lower than those found in semi-synthetic and synthetic media. No such decrease in activity was observed in the case of gramicidin S. In addition, the growth of *Escherichia coli* was inhibited by gramicidin S at a concentration of 10 μg/ml in the media. However, gramicidin S is known to be active against Gram-positive bacteria and less active toward Gram-negative bacteria such as *Escherichia coli*. Gramicidin S was inactive against *Escherichia coli* at the high concentration of 100 μg/ml under the authors' experimental conditions. Thus, careful consideration is required of differences in the experimental conditions used. Relative activities in the bacterial spectrum should be considered in order to compare results obtained in different laboratories.

Erlanger *et al.*[172] reported on the activities of linear pentapeptides and decapeptides related to gramicidin S. Although the intrinsic activities of these linear analogs were rather weak, a synergistic effect was observed when they were assayed in the presence of gramicidin S; the effect disappeared if two linear decapeptides were combined, as shown in Table 4.1. In regard to the time course of antibacterial action, gramicidin S produced immediate bacteriostasis, but linear peptides inhibited the growth of bacteria after the organisms had undergone several cell divisions. The mode of action of synthesized linear decapeptides, therefore, appears to be different from that of gramicidin S. In this connection, Erlanger *et al.* expressed doubt as to whether copolymers such as poly-(Orn, Leu) really react with microorganisms in the same way as gramicidin S.

Makisumi *et al.*[174,175,177] and Matsuura *et al.*[119,120] synthesized similar penta-, deca-, and eicosapeptides. Weak activities of these peptides again suggest that the mode of action of the synthetic linear peptides differs from that of gramicidin S, as shown in Tables 4.2. and 4.4.

6.2 Antibacterial Action of Gramicidin S

Several interesting reports were published by Otani's group dealing with the interaction of gramicidin S and the cell surfaces of microorganisms. Uemura treated sensitive organisms, *Bacillus subtilis* and *Staphylococcus aureus*, with ^{14}C-labeled gramicidin S.[278] After incubation, the bacterial cells were disrupted and fractionated by differential centrifugation. In case of the sensitive bacteria, some 80–88 % of radioactive gramicidin S was adsorbed by the cells and some 80 % of the adsorbed peptides was recovered from ribonucleoprotein fraction (140,000 × *g* precipitate). In the case of resistant mutants, only 25–35% of gramicidin S was adsorbed by the cells and some 60–67% of adsorbed activity was found in the cell-wall fractions (14,800 × *g*). Hashimoto obtained consistent results by the use of the mono-dansyl derivative of gramicidin S.[209] Numerous fluorescent granules were observed with a fluorescence microscope in treated cells of the sensitive *Bacillus megathelium*, but no such granules were observed in resistant strains of the same species. After mechanical disintegration, this fluorescent portion was collected by centrifugation at 100,000 × *g* for 15 min in the small particle fractions.

Miki observed that gramicidin S-sensitive strains of *Micrococcus pyogenes* released an acid-soluble nucleotide fraction even on short incubation for 1–20 min at 0° C.[279] Hiramatsu reported that gramicidin S lysed protoplasts of the sensitive bacteria, *Micrococcus lysodeikticus*, at a concentration of 10 μg/m*l*, while the protoplasts of the non-sensitive bacteria *Escherichia coli* were lysed slightly, and the protoplasts of gramicidin S-producing strains of *Bacillus brevis* were not affected by the antibiotic even at the high concentration of 100 μg/m*l*.[280]

Apart from studies on membranes of microorganisms, studies on a model membrane system were attempted by Pache *et al.*[281] Gramicidin S or polymyxin B was mixed with lecithin dispersed in water. When the mixture of gramicidin S and lecithin was heated for several minutes, the solution became optically clear. The solution formed a gel upon addition of KCl. The sensitivity of the gramicidin S-lecithin mixture to the addition of ions is an indication that the interaction is electrostatic in nature, in contrast to the hydrophobic interaction between polymyxin B and lecithin. A fatty acid tail in polymyxin seems to penetrate into the lipid chain layer, but such an interaction could not be observed for gramicidin S. In any case, gramicidin S solubilizes the lecithin layer, the resulting particles having a molecular weight below 100,000. They stated that the minimum inhibitory concentration of gramicidin S (10–100 μg/m*l*) corresponds well to the ratio of the antibiotic to lipid necessary for interaction in the model system. These results may not be directly related to the antibacterial action on bacterial membranes, because these membranes do not contain lecithin.[281] However, similar negatively charged phospholipids should be present in bacterial membrane components, so electrostatic interaction may well be the first step in the antibacterial interaction.

The importance of the subsequent hydrophobic interaction has been suggested by Kaprel'yantis *et al.*[206] They reported that gramicidin S increases the permeability of the protoplast membranes of *Micrococcus lysodeikticus* and finally leads to lysis, which occurs more rapidly at 4° C than at 22–37° C. Hydrophobic interactions seem to participate in the disturbance of the osmotic barrier of the membrane. Inhibitory activity by gramicidin S on membrane enzymes such as malate dehydrogenase was enhanced by washing the gramicidin S-treated membrane with NaCl solution. These findings also indicate that hydrophobic interactions occur in addition to ionic interactions. Of course, ionic interactions are probably the key to the mechanism of action of gramicidin S; the peptide is assumed to form an ionic linkage with negatively charged phospholipids in the membrane. However, 1,4-diaminobutane and 1,5-diaminopentane, with distances between their amino groups close to that of gramicidin S, do not have any inhibitory effect on the respiratory enzymes. Simple ionic interaction cannot explain the final antibiotic effect of gramicidin S. Kaprel'yantis *et al.* proposed that gramicidin S causes phase separation of the negatively charged phospholipids from other classes of lipids, and leads to disturbance of the barrier properties. In addition, gramicidin S interacts with the membrane proteins, and prevents their redistribution. As a result, membrane proteins of the respiratory chain are surrounded by alkyl chains with restricted motion, which affects their catalytic activities.

All of these observations support the view that gramicidin S permeates the outer membrane of the cell and interacts with the inner membrane of sensitive organisms, altering the permeabilty of the membranes and releasing intracellular components such as lipids and nucleoproteins. In contrast, gramicidin S cannot permeate the cell membranes of resistant bacteria.

If gramicidin S forms particulates of lipid-peptide complex, how can we demonstrate this? As an approach to this problem, we suggest the use of affinity chromatography. Particles containing the supposed receptor may be isolated from a solution of solubilized membrane obtained from the sensitive organisms through a column containing gramicidin S, bound to an insoluble carrier.

Morphological changes in the membranes of sensitive organisms were studied by electron microscopy. When polymyxin B[282)] and EM-49[283)] reacted with the membranes of sensitive bacteria, blebs and protrusions emerged from the cell membranes. Release of phospholipids and membrane particles was also observed. Similar studies should produce further progress in this field.

6.3 Antibacterial and Other Actions of Tyrocidines and Linear Gramicidins

6.3.1 ANTIBACTERIAL AND ION-TRANSPORT ACTIVITIES OF TYROCIDINES AND GRAMICIDIN A

Natural tyrocidines and linear gramicidins are produced by *Bacillus brevis* as a mixture named tyrothricin. The antibacterial activities of tyrocidines and analogs of tyrocidine A are summarized in Table 4.8. The activities are gener-

TABLE 6.2 Antibacterial Activities of Gramicidin A and Its Shortened Analogs (Source: ref. 180. Reproduced by kind permission of North-Holland Pub. Co.)

Compound†	Minimum inhibitory concentration (μg/ml)		
	S. aureus	*B. subtilis*	*E. coli* B
1	0.05–0.1	1–2	>10
2	1–2	2–4	>10
3	0.07–0.1	1	>10
4	>10	>10	>10
5–9	>25	>25	>25

† The structure of each compound is shown below, e.g. **2** is the formyl tridecapeptide ethanolamide with the Val^7–D-Val^8 dipeptidyl fragment deleted.

```
    1                   5                        10                       15
1 CHO–Val–Gly–Ala–D-Leu–Ala–D-Val–Val–D-Val–Trp–D-Leu–Trp–D-Leu–Trp–D-Leu–Trp–EA
2 CHO–Val–Gly–Ala–D-Leu–Ala–D-Val————————————Trp–D-Leu–Trp–D-Leu–Trp–D-Leu–Trp–EA
3 CHO–Val–Gly–Ala–D-Leu–Ala–D-Val–Val–D-Val————————————Trp–D-Leu–Trp–D-Leu–Trp–EA
4 CHO–Val–Gly———————————————————Val–D-Val–Trp–D-Leu–Trp–D-Leu–Trp–D-Leu–Trp–EA
5 CHO–Val–Gly–Ala–D-Leu–Ala————————————————————D-Leu–Trp–D-Leu–Trp–D-Leu–Trp–EA
6 CHO–Val–Gly–Ala–D-Leu–Ala–D-Val–Val–D-Val————————————————————————Trp–D-Leu–Trp–EA
7 CHO–Val–Gly—————————————————————————————Trp–D-Leu–Trp–D-Leu–Trp–D-Leu–Trp–EA
8 CHO–Val–Gly–Ala–D-Leu–Ala————————————————————————————————D-Leu–Trp–D-Leu–Trp–EA
9 CHO–Val–Gly——————————————————————————————————————————————Trp–D-Leu–Trp–D-Leu–Trp–EA
```

ally lower than that of gramicidin S. Table 6.2 shows the antibacterial activities of gramicidin A and of eight analogs with shorter chains, all of which were synthesized by Miroshnikov *et al.*[180] Gramicidin A is strongly active, the thirteen-membered analogs are highly potent, although somewhat weaker, and further contraction of the chain essentially causes inactivation. Harold and Baarda observed that growth inhibition of *Staphylococcus aureus* by gramicidin A was reversed by the addition of excess K^+ and proposed that the antibiotic activity of gramicidin A might involve its ion-carrying activity for K^+.[284] In this connection, it is noteworthy that the decrease in antibacterial activity observed for shortened analogs by Miroshnikov *et al.* was paralleled by a distinct decrease in the intensity of IR spectra in the 3290 cm^{-1} region. This finding suggests that the antibacterial activity is related to an intramolecularly hydrogen-bonded conformation forming an ion tunnel.

Linear gramicidins are known to increase the ionic permeability of biological and artificial membranes, forming ion tunnels and allowing the passage of ions through the membrane without ion-selectivity. The ion-carrying activity of tyrocidines is very weak, if any, compared to gramicidins.[285] Desformyl-gramicidin A, synthesized by Urry *et al.*, showed no ion-transport activity, nor competitive inhibition against gramicidin A action. However, dimers linked by a malonamide group showed high ion-carrying activity. On the basis of this and other NMR and CD experiments, Urry *et al.* proposed a π_{LD} left-handed helix structure, forming head-to-head dimers, with 6.3 residues per turn.[50] On the other hand, Veatch *et al.* isolated four interconvertible conformational species of gramicidin dimers by silica gel chromatography, and suggested these species to be a new family of double helices, two of which are parallel, while one of the other two is antiparallel.[286] Precise determination of the conformations of these dimers seems to be difficult in biological media because of their poor solubility.

6.3.2 ACTIVITIES OF TYROCIDINES AND GRAMICIDINS RELATED TO SPORULATION

Recently, several peptide antibiotics, such as polymyxin B,[287] and actinomycin D,[288] have been shown to inhibit the growth of the producing microorganisms and to induce sporulation. The natural role of peptide antibiotics may be related to sporulation, because, like other secondary metabolites, peptide antibiotics are produced when conditions are unfavorable for vegetative growth.[289] Sarkar and Paulus suggested a possible regulatory role of tyrothricin in the life cycle of *Bacillus brevis* ATCC 8185, an organism producing the antibiotic peptides.[289] They found that the addition of tyrothricin to an exponentially growing culture of *Bacillus brevis* ATCC 8185 at a concentration corresponding to that produced during sporulation (100 μg/m*l*) led to 30–40% inhibition of growth. A similar inhibition was observed in incorporation of ^{14}C-uracil, but incorporations of ^{3}H-thymidine and ^{14}C-lysine were not inhibited. Growth inhibition by tryothricin may be due to an effect on RNA synthesis. Gramicidin S only weakly inhibited RNA synthesis in spite of its structural similarity to tyrocidines, as shown in Table 6.3. They suggested that the peptide antibiotic functions in the regulation of gene transcription during the transition from vegetative growth to sporulation, and that the inhibition of RNA polymerase by the antibiotic is probably not due to its surface-active nature, because gramicidin S is a much stronger surfactant than linear gramicidins.

TABLE 6.3 Inhibition of RNA Polymerase by Peptide Antibiotics (Source: ref. 289. Reproduced by kind permission of Macmillan Journal Ltd.)

Peptide	% Inhibition by peptide	
	5 μg/m*l*	50 μg/m*l*
Tyrothricin	40	96
Tyrocidines	44	100
Linear gramicidin	50	90
Gramicidin S	—	23

In the next study, Mukherjee and Paulus showed that gramicidin-negative mutants of *Bacillus brevis* cannot form normal spores without addition of the antibiotic.[290] Furthermore, Sarkar *et al.* suggested the role of linear gramicidin in the transition from vegetative growth to sporulation to be selective inhibition of purified RNA polymerase.[291] Gramicidin inhibits the transcription of DNA of *Bacillus brevis* but does not inhibit it when RNA synthesis is inhibited 75% by the addition of low levels of actinomycin D (0.15 μg/m*l*). Gramicidin may selectively inhibit the transcription of some gene(s), but not all. In consequence, they suggested that inhibition of transcription by gramicidin occurs at the level of promotor recognition. Gramicidin may promote the dissociation of the "open" complex between RNA polymerase and template DNA. Since this inhibition was observed in a highly purified system, free from membrane fragments, the inhibiting activity must be distinct from the ionophoretic activity, which seems to be related to the antibiotic action as shown in Section 6.3.1.

Schazschneider *et al.* examined the effects of tyrocidines and linear gramicidins separately, and observed that tyrocidines interact with DNA to form a complex. This complex formation was detected by Sephadex G-25 chromatography and nitrocellulose filtration.[292] The fact that C-G rich regions in DNA exhibit intense hypochromicity in the presence of tyrocidine suggests that C-G may be the interacting site.[293] They further found that tyrocidines inhibit RNA synthesis, but linear gramicidins overcome this inhibition at least partly by weakening the binding of tyrocidines to DNA in the complex.[294] Although gramicidin itself inhibits transcription *in vitro*, the inhibition is not due to complex formation with DNA. When similar experiments were done *in vivo*, tyrocidine inhibited RNA synthesis and led to a cessation of growth of exponentially growing *Bacillus brevis*. In contrast, gramicidin does not affect net RNA synthesis or cell growth, but does inhibit sporulation.[295] When the two peptides were added as a mixture, inhibition of RNA synthesis by tyrocidine was partially released by gramicidin.

6.3.3 RELATED PROBLEMS ON THE ACTION OF TYROCIDINES AND GRAMICIDINS

Various biological activities of tyrocidines and gramicidins have been observed separately. Early in 1966, gramicidin S, tyrocidines, and gramicidins were known to inhibit the ATP-induced reversal of swelling of eukariotic mitochondria caused by reduced glutathione and to act as uncouplers of oxidative phosphorylation.[296] Inhibition of RNA synthesis was also observed as mentioned above. However, we do not know exactly what is the essential function of these peptides in the producing cells. In the case of gramicidin S, mutants devoid of gramicidin S-producing activity were found to be oligosporogeneous.[297] However, another gramicidin S non-producing mutant was found to sporulate at a higher percentage than the wild-type cells.[297] Thus, the situation is not simple. As described earlier, the antibacterial behavior of enantiogramicidin S was shown to be similar to that of gramicidin S.[202] This result indicates that the mode of antibacterial action of gramicidin S is not enzymatic but simply physico-chemical in nature, e.g., involving ionic and hydrophobic interactions. If enantiotyrocidine A can be synthesized and its potential activity tested, this should provide useful information for the elucidation of the biological role of the peptide and related antibiotics.

Recently, Church *et al.* presented an interesting speculation on structural supplementarity between DNA and protein based on molecular model building.[298] In their model, double-stranded DNA and double stranded antiparallel β-sheet peptides bind with intermolecular hydrogen bonding between 3′-phosphate and peptide amide. The interaction between tyrocidine and DNA might be explained by postulating a similar model, although tyrocidines are known to inhibit not only the transcription of double-stranded DNA but also that of single-stranded DNA.

CHAPTER 7

Biosynthesis of Gramicidin S and Tyrocidines

The biosynthesis of peptide antibiotics such as gramicidin S and tyrocidines has proved to be quite different from that of usual proteins which are produced by the ribosomal RNA-dependent mechanism. Recent investigations have demonstrated that the biosynthesis of these peptides is due to a series of reactions by nucleic acid-independent multienzymes, and biosynthetic studies are now one of the most interesting problems in the field of biochemistry. However, as this book emphasizes the chemical aspects of cyclic peptides, we will describe the biosynthesis of gramicidin S and tyrocidines only in outline. There are several articles and reviews on the biosynthesis of oligopeptides. Among them, the reviews by Kurahashi[299)] and Lipmann[300)] can be recommended.

7.1 Biosynthesis of Gramicidin S

7.1.1 ENZYMES CONCERNED IN GRAMICIDIN S SYNTHESIS

Early in the 1960's, studies on the biosynthetic mechanism of peptide antibiotics were started as simple models of protein biosynthesis. However, various data indicated that the biosynthesis of peptide antibiotics is dependent on a purely enzymatic process and is similar to that of fatty acids.[301)] Several oligopeptides which are possible intermediates in gramicidin S synthesis have been found. Tomino and Kurahashi observed that an incubation mixture of L-phenylalanine and L-proline with cell-free extracts from *Bacillus brevis* Nagano gave D-Phe–L-Pro anhydride, which might be derived from a linear dipeptide, D-Phe–L-Pro.[302)] When L-phenylalanine, L-valine and L-proline were used, a tripeptide, H–D-Phe–L-Pro–L-Val–OH, was formed. These results suggested that the formation of D-Phe–L-Pro was the first step leading to the biosynthesis of gramicidin S.[302] Moreover, the formations of H–D-Phe–L-Pro–L-Val–L-Orn–OH[303] and H–D-Phe–L-Pro–L-Val–L-Orn–R (in which R was unknown)[304] were also demonstrated in other incubation systems.

Otani *et al.* found that the conversion of L-phenylalanine to the D-isomer occurred during dipeptide formation, because the dipeptide always contained a D-phenylalanine residue in spite of the use of the L-isomer as s starting amino acid, and they also found that the rate of the incorporation of D-phenylalanine into the peptides was far less than when the L-isomer was used.[305] The structures of the peptides isolated in their experiments suggested that L-phenylalanine is the starting amino acid in gramicidin S biosynthesis and the peptide chain grows from the *N*-terminal to the *C*-terminal.

On the other hand, isolation and characterization of gramicidin S-synthesizing enzymes have been attempted by several groups.[301,306–8] So far, gramicidin S- and tyrocidine-synthesizing systems have been fractionated into two and three components, respectively. Their characteristics are summarized in Table 7.1. Tomino *et al.*[306] reported the isolation of two complementary fractions, light enzyme and heavy enzyme, from the gramicidin S-synthesizing system by Sephadex G-200 chromatography with 0.02 M triethylamine–HCl buffer (pH 7.4) containing 0.01 M $MgCl_2$. In preliminary studies, they observed ATP-PPi exchange activities dependent on L-leucine and L-valine in both fractions, whereas ATP-dependent racemization activity towards L-phenylalanine was present only in the light enzyme. The results suggested that the light enzyme activates L-phenylalanine and takes the initiative in peptide formation.[306]

Hydroxylapatite chromatography was also effective for the fractionation

TABLE 7.1 Some Characteristics of Gramicidin S and Tyrocidine Synthesizing Enzymes (Source: ref. 328. Reproduced by kind permission of Kyoritsu Shuppan Co.)

Enzyme system	Components of enzyme system	Amino acid activated by enzyme	Phosphopantetheine (mole) /component (mole)	MW
gramicidin S synthesizing system	LE†1	L- and D-Phe	0	100,000
	HE†2	L-Pro, L-Val, L-Orn, L-Leu	1	280,000
tyrocidine synthesizing system	LE†1	L- and D-Phe	0	100,000
	IE†3	L-Pro, L- and D-Phe	1	230,000
	HE†2	L-Asn, L-Gln, L-Tyr, L-Val, L-Orn, L-Leu	1	440,000

†1Light enzyme. †2Heavy enzyme. †3Intermediate enzyme.

of the gramicidin S-synthesizing system.[305] The light enzyme has a molecular weight of 100,000[307] and its activity is stimulated by thiol compounds such as dithiothreitol and 2-mercaptoethanol.[309] The rate of the formation of L-phenylalanine from the D-isomer catalyzed by the light enzyme was slower than that of D-phenylalanine from the L-isomer, and L-phenylalanine and D-phenylalanine were found to be present in a ratio of 3 to 7 when the reaction reached apparent equilibrium.[309]

After incubation of five radioactive amino acid constituents of gramicidin S with the gramicidin S-synthesizing system and ATP, enzyme protein was precipitated by treatment with trichloroacetic acid. Radioactivity in the precipitated protein could be removed by treatment with performic acid or 0.1 N KOH, indicating the possibility that the amino acids are bound to the enzyme through a thioester linkage.[310,311] Independently, Lipmann's group concluded that the linkage between the amino acids and enzyme protein was thioester in nature by demonstrating the cleavage of the linkage by alkali, hydroxylamine or sodium borohydride.[312] As shown in Table 7.1, the heavy enzyme activates L-proline, L-valine, L-ornithine and L-leucine, which are bound in the form of thioesters at separate reactive sites on the enzyme molecule.[313]

7.1.2 GROWTH OF THE PEPTIDE CHAIN

The direction of growth of the peptide chain and the formation of a thioester-bound nascent peptide chain have been studied by two groups. In the experiments of Lipmann's group,[312] light enzyme and heavy enzyme were incubated with ATP, Mg^{++} and various combinations of amino acids including ^{14}C-L-phenylalanine. After treatment of the trichloroacetic acid-precipitable protein isolated by Sephadex G-50 chromatography with alkali, radioactivity of phenylalanine released from the enzyme protein was counted. The amount of radioactivity increased only in the case of addition of the amino acids in the order L-phenylalanine, L-proline, L-valine and L-ornithine.[312] As shown in Table 7.2, the protein-bound intermediates were formed even when all five amino acids were presented in the incubation system, but the radioactivity of total protein-bound intermediates was much less than when L-leucine was omitted. On the other hand, gramicidin S was formed in a large amount. Moreover, they observed that the omission of L-proline cut down the amount of bound radioactivity to that of L-phenylalanine alone, in spite of the further addition of L-valine and L-ornithine.[312] Ljones *et al.*[310] also reported that protein-bound intermediates participated in gramicidin S synthesis by a pulse-chase experiment using labeled amino acids.

In order to identify the protein-bound intermediates, the materials were treated with hydroxylamine to yield peptidyl hydroxamates.[312] Electropho-

TABLE 7.2 Formation of Protein-bound Peptide Intermediates and Gramicidin S (Source: ref. 312. Reproduced by kind permission of Prof. F. Lipmann)

Added amino acid†	Radio-activity in Sephadex eluate (cpm)	Non-TCA-precipitable radio-activity (cpm)	TCA-precipitable radio-activity (cpm)	GS (cpm)
^{14}C-L-Phe	2,960	1,410	1,550	265
^{14}C-L-Phe,L-Pro	7,670	1,670	6,000	375
^{14}C-L-Phe,L-Pro,L-Val	11,500	1,460	10,000	1,530
^{14}C-L-Phe,L-Pro,L-Val,L-Orn	16,200	1,840	14,300	1,000
^{14}C-L-Phe,L-Pro,L-Val,L-Orn,L-Leu	3,530	1,500	2,030	25,400
^{14}C-L-Phe,———,L-Val,L-Orn	3,100	1,490	1,610	376

† 0.5 nmole of ^{14}C-L-Phe (0.2 μcurie), 10 nmole of other amino acids and 5 μg each of light enzyme and heavy enzyme were used.

resis of the hydroxamates revealed the presence of Phe–Pro anhydried and tri-, tetra- and pentapeptidyl hydroxamates, but there was no evidence for the presence of peptides between six and nine residues long in this system. From the results, Lipmann's group suggested the possibility that a dimeric reaction between two antiparallel pentapeptide units occurred to form gramicidin S.[312)]

A series of studies by Lipmann and his co-workers revealed that the participation of thioester intermediates in gramicidin S synthesis by the multienzyme system was quite similar to the two-carbon addition mechanism in fatty acid synthesis. Hence, Gevers *et al.* speculated that phosphopantetheine (**7.1**) might play an important role in gramicidin S synthesis.[312)] Soon after-

$$HS(CH_2)_2NHCO(CH_2)_2NHCOCH(OH){-}C(CH_3)_2CH_2OP(=O)(OH){-}OH$$

(**7.1**).

wards, Gilhuus-Moe *et al.* demonstrated that upon alkaline treatment of the two enzymes involved in gramicidin S synthesis, one molecule of pantothenic acid was released per molecule of the heavy enzyme (see Table 7.1).[301)] The function of phosphopantetheine is presumably to bring the growing intermediate to the site of the next thioester-linked activated amino acid on the surface of the enzyme molecule. Another detailed study was carried out by Kleinkauf *et al.*[314)] After L-phenylalanine and ^{14}C-L-proline had been incubated in the gramicidin S-synthesizing system, trichloroacetic acid-precipitable protein bound to the D-Phe–L-Pro moiety was isolated. Pepsin digestion of the D-Phe–L-Pro-bound protein followed by thin-layer chromatography showed that the position of radioactivity from D-Phe–L-Pro coincided with that of pantothenic acid, which was checked biologically.

On the strength of the above and other results, Kleinkauf *et al.* proposed a hypothesis for the role of the pantetheine in the heavy enzyme molecule during the growth of the peptide chain.[314] An "arm" of phosphopantetheine stretches from the center of the heavy enzyme molecule toward the amino acid-activating sites aligned on the surface of the enzyme. The thiol group of pantetheine takes up a peptide intermediate which is linked through a thioester linkage to the enzyme molecule. Then, the arm approaches the next reactive site of the same enzyme molecule and transfers the peptidyl moiety to the amino group of the aminoacyl thioester. By repetition of transpeptidation from one thiol group to another *via* the pantetheine arm, the peptide chain grows stepwise up to the pentapeptide level.

Therefore, light enzyme is not necessary for the growth of the peptide chain after the formation of D-Phe–L-Pro. The D-phenylalanine residue activated by light enzyme was shown to be transferred directly to the imino group of L-proline linked to the heavy enzyme to form the D-Phe–L-Pro linkage.[314]

7.1.3 FORMATION OF GRAMICIDIN S

One of the most interesting problems in gramicidin S synthesis is the cyclization mechanism of the peptide formed on the heavy enzyme. As described already, Gevers *et al.* proposed a dimeric reaction between two pentapeptides.[312] The ingenious work by Stoll *et al.* confirmed this hypothesis by using DL-azetidine-2-carboxylic acid **(7.2)** instead of L-proline in the

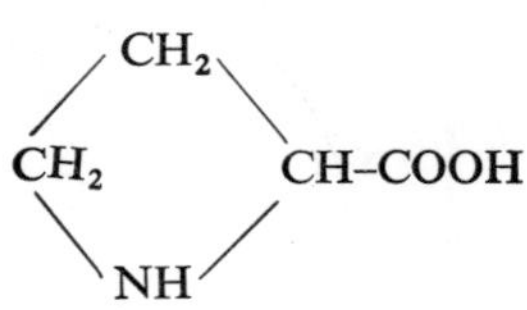

(7.2)

biosynthesis of gramicidin S as follows.[315] When two kinds of heavy enzymes with peptides containing L-azetidine-2-carboxylic acid and L-proline, respectively, are mixed, no hybrid decapeptide will be formed if the reaction is intramolecular, whereas the formation of 50% of the hybrid decapeptide containing one molecule of L-azetidine-2-carboxylic acid and one of L-proline as well as two other decapeptides is expected if the reaction is intermolecular,

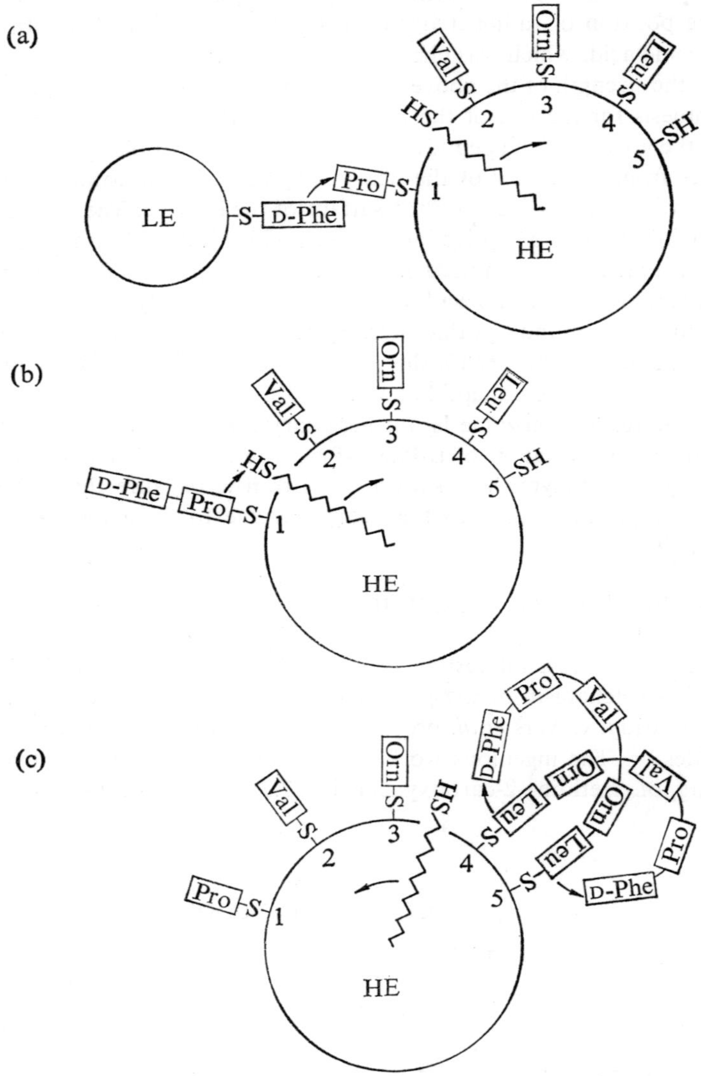

Fig. 7.1 Tentative scheme for gramicidin S biosynthesis proposed by Laland *et al.* Positions 1–4 are the SH sites and position 5 is the waiting site. The pantetheine arm is shown by a zigzag line. (Source: ref. 317. Reproduced by kind permission of Macmillan Journal Ltd.)

Two separate incubation mixtures of gramicidin S-synthesizing system containig ^{3}H-L-proline and ^{14}C-DL-azetidine-2-carboxylic acid, in addition to L-phenylanaine, L-valine and L-ornithine were combined and indubated with the constituent amino acids of gramicidin S and ATP. The labeled cyclic decapaptides formed were separated by thin-layer chromatography and counted by radioautography. The spot of gramicidin S contained 91% of the total ^{3}H label and was contaminated with a small amount of ^{14}C label. On the other hand, the spot of the analog with L-azetidine-2-carboxylic acid contained 95% of the ^{14}C label and 9% of the ^{3}H label, showing that the reaction was intramolecular.[315]

These interesting results suggest that another thiol group is required, to which the first pentapeptide formed is temporarily linked during the formation of the second pentapeptide. A thiol group near the leucine-activating site is now thought to be this "waiting" site.[316] The second pentapeptide can be brought into such close contact with the waiting pentapeptide that head-to-tail cyclization occurs. The failure to find peptides longer than the pentapeptide supports this concept of cyclic condensation.[312,315]

Based on the findings described above, Laland *et al.* published a tentative but attractive scheme for the biosynthesis of gramicidin S as illustrated in Fig. 7.1.[317] In Fig. 7.1a, a D-phenylalanine residue is transferred from the light enzyme to the imino group of L-proline linked to the heavy enzyme. This reaction occurs without participation of the pantetheine arm. The dipeptide formed is transferred to the thiol group of the pantetheine (Fig. 7.1b) and then forms the tripeptide chain by transferring to the amino group of valine. The same reaction is repeated up to the formation of the pentapeptide, which must then be transferred to a waiting site. After another pentapeptide has been formed in the leucine site, cyclization of the two pentapeptides on the same enzyme molecule takes place to give gramicidin S (Fig. 7.1c). However, Roskoski *et al.* have described another possibility, that the activated pentapeptides formed on two separate enzyme molecules yield gramicidin S as shown in Fig. 7.2.[313] These inconsistent results indicate that the problem of the cyclization mechanism in gramicidin S still requires further study.

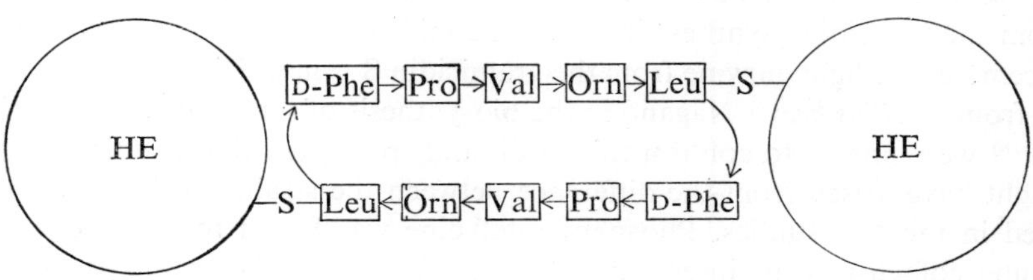

Fig. 7.2 Cyclization mechanism involving two pentapeptides on separate enzyme molecules according to Roskoski *et al.*

Spurred by the idea of the formation of gramicidin S by a dimeric cyclization reaction *in vivo*, several cyclizations of a pentapeptide (**7.3**) having a *C*-terminal L-leucine residue have been applied in chemical syntheses of grami-

H–D-Phe–L-Pro–L-Val–L-Orn(Z)–L-Leu–ONSu

(**7.3**)

cidin S,[70,71] but considerable amounts of cyclic pentapeptide were formed in addition to gramicidin S,[70] as described in Section 3.1.1. The apparent discrepancy may be based on the difference in the conditions of cyclization. *In vivo*, the precursor having two free amino groups is activated as a thiol ester, and dimeric cyclization is carried out in an aqueous environment, where the enzyme protein may also have some favorable effect in the formation of gramicidin S from the precursor. On the other hand, in organic synthesis the δ-amino group of the ornithine residue in the precursor was protected and cyclization was carried out by the active ester method or the azide method in an organic solvent.

7.2 Biosynthesis of Tyrocidines

It is not long since investigations of the biosynthesis of tyrocidines began. Lipmann's group[318,319] and Kurahashi's group[320,321] reported at almost the same time the isolation and characterization of enzymes involved in tyrocidine biosynthesis. In the study of the former group, an enzyme system from *Bacillus brevis* could be separated by Sephadex G-200 gel filtration into three components, named light enzyme (molecular weight 100,000), intermediate enzyme (230,000) and heavy enzyme (440,000) as shown in Table 7.1. The light enzyme and intermediate enzyme activate L-phenylalanine and L-proline, respectively, and the heavy enzyme activates the remaining constituent amino acids of tyrocidine. The light enzymes of tyrocidine and gramicidin S are similar in size and function.[322]

Kurahashi and his co-workers[320,322] demonstrated that the light enzyme from the tyrocidine-synthesizing system could be replaced by phenylalanine racemase (the light enzyme from the gramicidin S-synthesizing system) purified from *Bacillus brevis* Nagano in the biosynthesis of tyrocidine. Roskoski *et al.*[318] were unable to confirm this result and speculated that the discrepancy might have arisen from the difference between the strains of *Bacillus brevis* used in the two studies. Phosphopantetheine was found to be an enzyme-bound cofactor in the intermediate enzyme and in the heavy enzyme of the three enzymes,[314,319,323] and acted as a peptide carrier,[314] as in gramicidin S biosynthesis.

The light enzyme activates L-phenylalanine and then growth of the peptide chain occurs.[319–21] A skillful study by Roskoski *et al.*[319] clarified the mechanism of growth of the peptide chain using radioactive amino acids. The results are shown in Table 7.3. Enzyme-bound peptide intermediates up to the linear decapeptide could be isolated from the system by precipitation with trichloroacetic acid followed by alkali treatment, and were identified by thin-layer chromatography. Table 7.3 indicates that the requisite amino acids must be added in a definite order for peptide synthesis. For example, omission of a certain amino acid (entries 8, 9 and 10 in Table 7.3) cut down the growth of the peptide chain, even if the succeeding amino acids were added. These results show that the direction of chain elongation is the same as that of gramicidin S.

TABLE 7.3 Formation of Enzyme-bound Nascent Peptide Chains (Source: ref. 319. Reproduced by kind permission of The American Chemical Society)

Entry	Added amino acid†	Bound radioactivity (pmoles)
1	^{14}C-L-Phe	25
2	^{14}C-L-Phe,L-Pro	49
3	^{14}C-L-Phe,L-Pro,L-Asn	67
4	^{14}C-L-Phe,L-Pro,L-Asn,L-Gln	101
5	^{14}C-L-Phe,L-Pro,L-Asn,L-Gln,L-Val	132
6	^{14}C-L-Phe,L-Pro,L-Asn,L-Gln,L-Val,L-Orn	148
7	^{14}C-L-Phe,L-Pro,L-Asn,L-Gln,L-Val,L-Orn,L-Leu	182
8	^{14}C-L-Phe,—,L-Asn,L-Gln,L-Val,L-Orn,L-Leu	25
9	^{14}C-L-Phe,L-Pro,—,L-Gln,L-Val,L-Orn,L-Leu	51
10	^{14}C-L-Phe,L-Pro,L-Asn,L-Gln,L-Val,—,L-Leu	120

† 6 nmole of ^{14}C-L-Phe (80 μcurie/μmole) and other amino acids, 40 μg of light enzyme, 100 μg of intermediate enzyme and 150 μg of heavy enzyme were used.

Fig. 7.3 shows the biosynthetic pathway of tyrocidine according to Roskoski *et al.*[319] The light enzyme with activated D-phenylalanine reacts with the imino group of L-proline on the intermediate enzyme to form D-Phe–L-Pro–thioenzyme. Thus, the growth of the peptide chain by peptidyl transfer through thioester-bound intermediates occurs until the decapeptidyl thioenzyme is formed. Finally, reaction between the carboxyl group activated as the thioester of L-leucine and the amino group of D-phenylalanine causes cyclization and release of the cyclic product, tyrocidine, from the surface of the heavy enzyme molecule.[319] As compared with that of gramicidin S, the rate of cyclization of the linear decapeptide precursor corresponding to tyrocidine is slow.

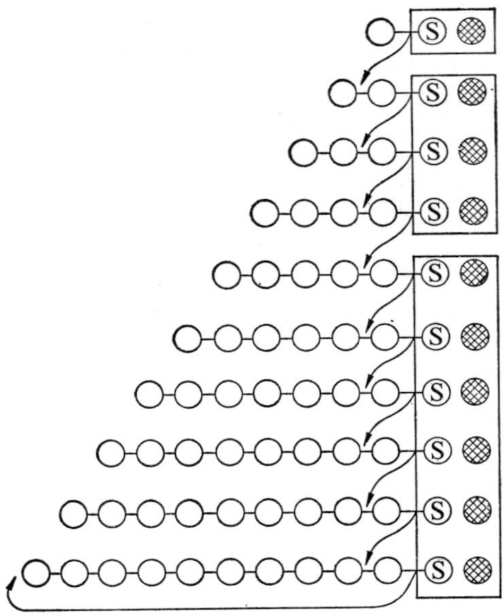

Fig. 7.3 Scheme for tyrocidine biosynthesis. ◍, individual sites in three enzyme fractions for the binding of aminoacyl adenylates; Ⓢ, SH groups for thioester binding of amino acids or peptides; ○, amino acid residues. (Source: ref. 319. Reproduced by kind permission of The American Chemical Society)

7.3 Biosynthesis of Gramicidin S Congeners and Tyrocidine Congeners

It has often been observed that many natural antibiotics are mixtures of congeners in which one or more amino acid residues in the peptide chain are replaced by structurally similar ones. Mach and Tatum speculated that simultaneous production of multiple antibiotics is due to the broad specificity of enzymes.[324)] Indeed, the formations of gramicidin S and tyrocidine analogs have been observed on the addition of a similar amino acid instead of a constituent amino acid of the antibiotics to an *in vivo* or *in vitro* system.

In vivo experiments by Winnick and Winnick[325)] showed that gramicidin S analogs were produced to a small extent when a large amount of DL-norleucine or DL-*p*-fluorophenylalanine was supplied instead of L-leucine or L-phenylalanine. As described in Section 7.1.3, Stoll *et al.*[315)] obtained a gramicidin S analog containing L-azetidine-2-carboxylic acid instead of L-proline using an *in vitro* system, while no incorporation of other imino acids such as L-hydroxyproline and L-pipecolic acid was observed. D-Phenylalanine and L-valine residues in gramicidin S can be replaced by structurally related aromatic and branched amino acids, respectively.[326]

The biosynthesis of tyrocidine congeners is influenced by the amino acid concentration in the medium as shown in Table 7.4. Tyrocidines A, B and C are produced in a ratio of 1:3:7 under usual *in vivo* conditions.[324] Mach and Tatum found that tyrocidine A was exclusively formed when L-phenylalanine was added to the medium, while tyrocidine D was mainly formed when L-tryptophan was used.[324] L-Leucine and L-valine residues in tyrocidine can also be replaced by branched amino acids such as L-alloisoleucine and L-isoleucine. L-Proline, however, is hardly replaced by L-pipecolic adid.[327] Similar results were later obtained *in vitro* by Fujikawa *et al.*[33] as shown in Table 7.4. They also obtained lysine-tyrocidine A, isoleucine-tyrocidine A and tyrocidine E as described in Section 2.2.[33]

TABLE 7.4 Formation of Tyrocidine Congeners under Various Conditions

System	Conditions	Product	Reference
in vivo	minimal medium	Tyrocidine A, B, C (1:3:7)	324
	+ L-Phe	Tyrocidine A (major)	324
		Tyrocidine B, C (minor)	
	+ L-Trp	Tyrocidine D (major)	324
		Tyrocidine A, B, C (minor)	
	+ L-Phe, L-Trp	Tyrocidine A, B, C, D	324
in vitro	standard mixture	Tyrocidine D (major)	33
		Tyrocidine C (minor)	
		Tyrocidine A, B (not detectable)	
	− L-Tyr	Tyrocidine D	33
	− L-Trp	Tyrocidine A	33
	− L-Trp (small amount)	Tyrocidine A, B, C	33
	− L-Tyr, L-Trp	Tyrocidine E	33

Table 7.5 shows the specificity in the incorporation of various amino acids other than the constituent amino acids of gramicidin S or tyrocidines under *in vivo* or *in vitro* conditions. In the case of a DL-amino acid as an amino acid source, it is not known whether both isomers are equally activated by the enzyme. Lipmann *et al.*, however, observed that none of the D-analogs of the L-amino acids found in the antibiotics entered into peptide biosynthesis.[326] The specificity of the enzymes in the tyrocidine-synthesizing system seems to be lower than in the gramicidin S-synthesizing system.[328]

Although a strict comparison of the degree of amino acid incorporation

TABLE 7.5 Specificity of Amino Acid Incorporation in Gramicidin S and Tyrocidine Congener Biosynthesis

Antibiotic	System	Original amino acid	Incorporated amino acid	Non-incorporated amino acid	Ref.
Gramicidin S	*in vivo*	L-Val		DL-aminoisobutyric acid	325
		L-Orn		DL-diaminobutyric acid DL-Lys	325
		L-Leu	L-norleucine		325
		D-Phe	DL-thienylalanine DL-*p*-fluorophenyl-alanine		325
		L-Pro		DL-pipecolic acid L-Hyp	325
	in vitro	L-Val	L-Ile L-aIle L-norvaline	D-Val D-Ile	326
		L-Orn		L-Lys L-Arg D-Orn	326
		L-Leu	L-Ile L-aIle L-norleucine	D-Leu D-Ile	326
		D-Phe	DL-thienylalanine L-*p*-fluorophenyl-alanine	L-Tyr L-Trp DL-*β*-threo-phenylserine	326
		L-Pro	DL-azetidine-2-carboxylic acid	L-Hyp L-pipecolic acid *trans*-3-methylproline *trans*-4-fluoroproline	315
Tyrocidine	*in vivo*	L-Leu L-Val	L-Ile L-aIle		327
		D-Phe	DL-thienylalanine		327
		L-Pro	L-pipecolic acid		327
	in vitro	L-Val	DL-norvaline		33
		L-Orn	L-Lys L-diamino-butyric acid	L-diaminopimelic acid	33
		L-Leu	L-Ile DL-norleucine		33
		D-Phe	DL-thienylalanine DL-*p*-fluoro-phenylalanine	DL-*β*-threo-phenyl-serine	33
		L-Pro	L-Hyp		33
		L-Trp	DL-5-methyltryptophan DL-*p*-fluorotryptophan		33

is difficult because of the use of various biosynthetic conditions, Katz and Demain[329] have indicated that amino acid replacement is often found at the site of an aromatic or branched chain amino acid residue in peptide antibiotic congeners. Ishihara and Shimura speculated that rigid structure of the peptide backbone rather than the side chains of amino acid residues might be required for cyclization, because the replacement of a given amino acid residue by similar amino acids is possible.[328]

CHAPTER **8**

Synthetic Aspects of AM-toxins

During recent years, some cyclic peptides and cyclodepsipeptides possessing small rings and interesting structural and biological features have been found. A fungus, *Alternaria tenuis* Nees, synthesizes a cyclic tetrapeptide, tentoxin, which causes chlorosis in germinating seedlings.[330] Koncewicz *et al.* reported the amino acid sequence of tentoxin,[331] but soon afterwards Mayer *et al.* revised the structure of tentoxin proposed by Koncewicz's group on the basis of chemical and X-ray analyses, reporting it to be as shown in Fig. 8.1.[332,333]

Hirota *et al*[334–6] isolated three metabolites designated Cyl-1, -2 and -3 as plant growth regulators from *Cylindrocladium scoparium* Morgan, a phytopathogenic fungus, and elucidated the primary structure of Cyl-2, which has marked inhibitory activity on the root growth of lettuce seedlings, as cyclo(–Aoe–D-Tyr(Me)–L-Ile–L-Pip–), as shown in Fig. 8.1. Aoe indicates a 2-amino-8-oxo-9,10-epoxydecanoic acid residue and Pip indicates a pipecolic acid residue. However the absolute configuration of Aoe remains unclear. In 1974, Closse and Huguenin described the isolation, structure and activity of chlamydocin from metabolites of *Diheterospore chlamydosporia*. It possesses pronounced cytostatic activity.[337] The primary structure of chlamydocin is shown in Fig. 8.1. Interestingly, Cyl-2 and chlamydocin contain the same new amino acid, Aoe, and their amino acid compositions are similar.

Recently, cyclodepsipeptides named AM-toxins were isolated from *Alternaria mali* and have been studied physico- and organochemically to investigate their structure-activity relationships. In this chapter, we will describe the results of recent investigations of AM-toxins, including our results on the organic synthesis of AM-toxins and analogs. There is a recent review on the chemistry of peptides containing a dehydroamino acid, including AM-toxins.[338]

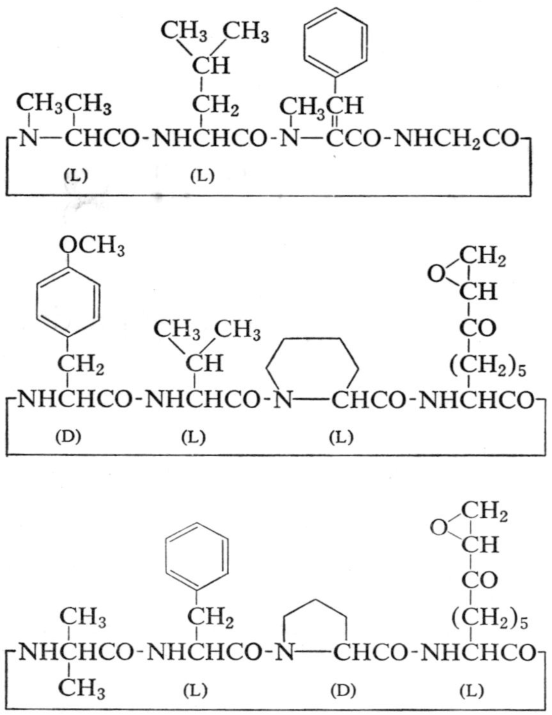

Tentoxin

Cyl-2

Chlamydocin

Fig. 8.1 Primary structures of cyclic peptides with a 12-membered ring.

8.1 Isolation and Primary Structures of AM-toxins

8.1.1 ISOLATION AND PROPERTIES OF AM-TOXINS

Alternaria blotch, a destructive disease of apples, is caused by a pathogenic strain of *Alternaria mali* Roberts; apple cultivars such as Indo and Delicious are known to be susceptible, while Jonathan is resistant to the disease. Sawamura observed in 1966 that the culture filtrate of the fungus and the pathogen caused the same necrotic brown spots on leaves of susceptible apples. Hence he speculated that the fungus might excrete a host-specific toxin.[339)]

Several years later, Okuno and his co-workers succeeded in the isolation of the host-specific toxins as follows.[340,341)] A culture broth from *Alternaria mali* Roberts (strain AKI-3) was filtered and the filtrate was extracted with chloroform. Evaporation of the extract gave an oily material, which was further purified by silica gel chromatography with chloroform–ethanol to give crude crystals. The pure compound obtained after recrystallization from chloroform–ethanol gave a melting point of 220–222° C and showed one spot on silica gel thin-layer chromatography. They found that the compound caused brown spots on leaves of Indo cultivar at a concentration of 1 μg/ml, while the same solution did not affect the leaves of the resistant Jonathan.[340,341]

They named the toxin alternariolide, and reported some properties of the toxin: melting point, 220–222° C; $[\alpha]_D^{18} -87.2°$ ($CHCl_3$). The molecular formula was established by mass spectrometry. The IR spectrum revealed the presence of secondary amide, ester, phenyl, methoxy and exomethylene groups.[342]

Different methods were applied for the isolation of the toxins by another Japanese group.[343–7] Okuno *et al.* isolated one compound, while Ueno *et al.* isolated three compounds, as shown in Fig. 8.2,[343–7] and Kohmoto *et al.* have suggested the presence of other toxins.[348] The procedure of isolation of Ueno *et al.*[343–6] was as follows. The filtrate (64 l) from the culture medium of *Alternaria mali* (strain I-716) was extracted with ethyl acetate, while the mycelial mat was extracted with acetone. After evaporation down of the extract, the residue was again extracted with ethyl acetate. The extracts from the filtrate and the mycelial mat were evaporated down to give an oily mass (11 g), which was dissolved in water. The aqueous solution was extracted successively with n-hexane, cyclohexane, toluene, dichloromethane and ethyl acetate. They found in this procedure that the dichloromethane extract possessed the most potent activity for inducing necrosis, and they then chromato-

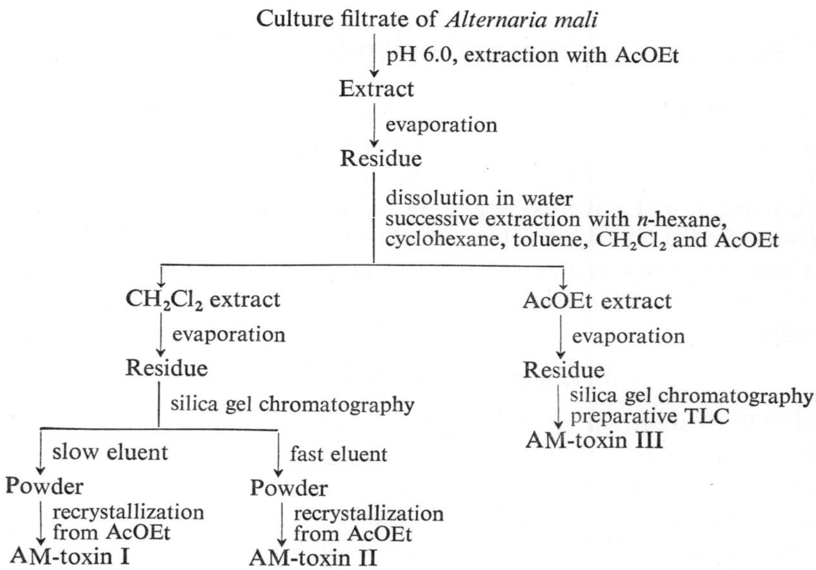

Fig. 8.2 Isolation of AM-toxins.

graphed the brown residue (4 g) obtained on a silica gel column using stepwise mixing ratio changes of chloroform–ethyl acetate mixture. Two active principles, named AM-toxins I and II, were found in the eluate with chloroform–ethyl acetate mixture (7:3–6:4 by volume). AM-toxin II was eluted first and then AM-toxin I. Recrystallization of each compound from ethyl acetate afforded pure AM-toxin I (89 mg) and AM-toxin II (50 mg).[343–6)]

The R_f values of AM-toxins I and II on a thin-layer chromatogram with cyclohexane–ethyl acetate–1-propanol (15:10:1 by volume) were 0.25 and 0.30, respectively. The UV spectrum of AM-toxin I gave an absorption maximum of $\varepsilon = 20{,}600$ at 224 nm in methanol and the IR showed the presence of peptide and ester bonds. The molecular formula as determined by elemental analysis was found to be identical with that derived from mass spectrometry. AM-toxin I had a melting point of 204–208° C and $[\alpha]_D^{19}$ −79° ($CHCl_3$).[346)]

The molecular extinction coefficient (ε) in the UV spectrum of AM-toxin II was small compared with that of AM-toxin I, while the IR spectra of AM-toxins I and II were very similar. AM-toxin II had a melting point of 212–4° C and $[\alpha]_D^{27}$ −14.2° (dimethylsulfoxide).[346)] Ueno *et al.*[347)] further succeeded in the isolation of a third toxic principle, AM-toxin III, after the work on the isolation of AM-toxins I and II. AM-toxin III was also isolated from the ethyl acetate extract by a similar procedure. The mass spectrum of AM-toxin III gave unequivocal evidence for the molecular formula, and the melting point was 228° C.[347)]

8.1.2 PRIMARY STRUCTURES OF AM-TOXINS

Okuno *et al.* characterized alternariolide as a cyclodepsipeptide containing a dehydroalanine residue on the basis of spectroscopic evidence[340,342)] and then determined the configurations of the constituent amino acids except for dehydroalanine and a hydroxy acid to be L-form.[349)] At the same time, Ueno *et al.* reported the primary structures of AM-toxins I and II, determined mainly by mass spectrometry and NMR.[345,346)] As a result, it was found that AM-toxin I was same as alternariolide; hereafter the name AM-toxin I will be used throughout.

A. Primary Structure of AM-toxin I

The PMR spectrum of AM-toxin I is shown in Fig. 8.3a. As already described by Okuno *et al.*[342)] the spectrum and the molecular formula indicated that it was a peptide containing a lactone and a *p*-methoxyphenyl group. On the other hand, as shown in Fig. 8.3b, the PMR spectrum of AM-toxin II indicated that it was different from AM-toxin I only in respect of the presence of a phenyl group instead of the *p*-methoxyphenyl group.

Ueno *et al.* confirmed the presence of a lactone linkage in AM-toxin I by

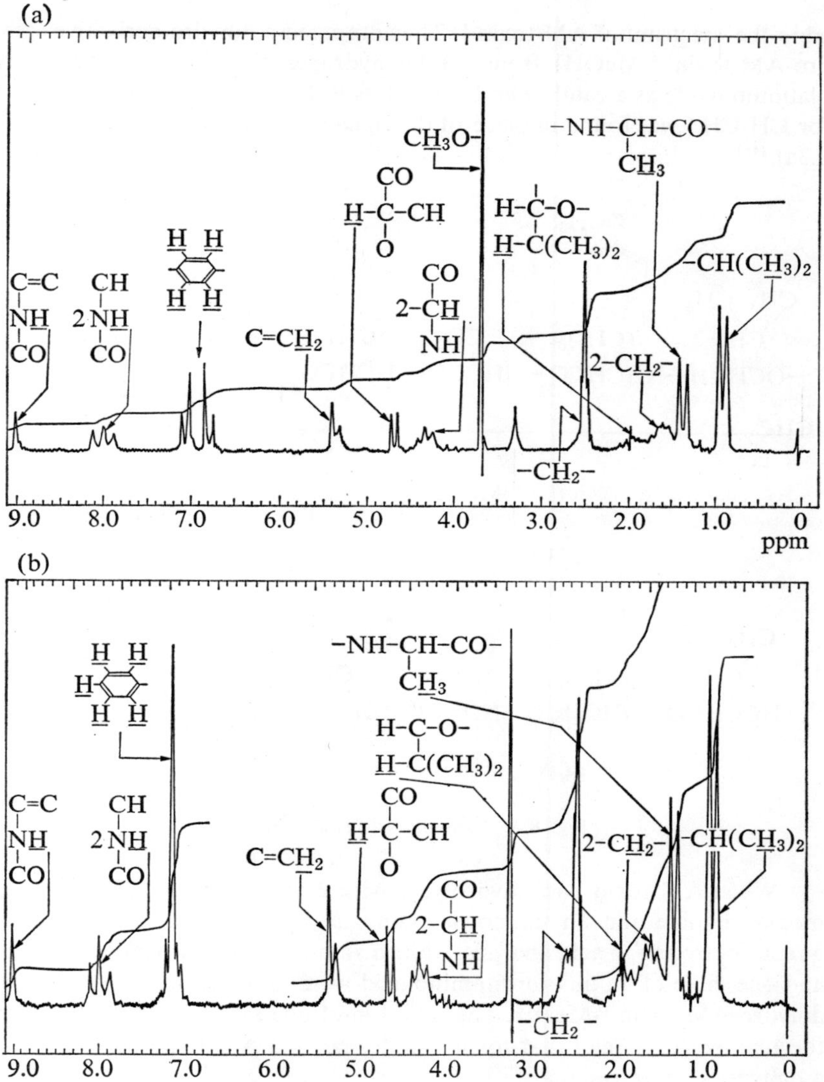

Fig. 8.3 NMR spectra of AM-toxin I (a) and AM-toxin II (b).

methanolysis as follows.[345,346] By treatment with methanol overnight at room temperature, AM-toxin I was readily converted to the corresponding hydroxy-carboxy ester, named AM-toxin I-MeOH (**8.1**), as shown in Fig. 8.4. High resolution mass spectrometry showed a molecular ion consistent with AM-toxin I + MeOH and its PMR spectrum showed two signals, a 3H singlet of $-OCH_3$ and a 1H broad signal of $-OH$ in addition to all the signals observed in the spectrum of AM-toxin I. They prepared a dihydro derivative (dihydro-AM-toxin I-MeOH) from **8.1** by hydrogenation in acetic acid using platinum oxide as a catalyst and studied its PMR spectrum. A doublet signal for $CH_3CH\langle$ was found in place of the signals for $CH_2 = C\langle$ in **8.1** (see Fig. 8.3a).[346]

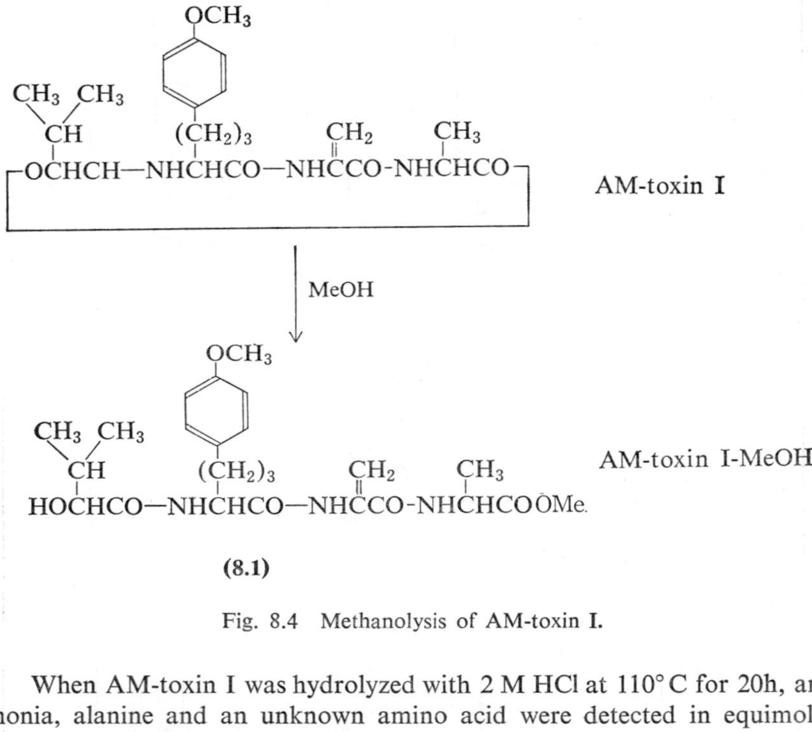

(8.1)

Fig. 8.4 Methanolysis of AM-toxin I.

When AM-toxin I was hydrolyzed with 2 M HCl at 110° C for 20h, ammonia, alanine and an unknown amino acid were detected in equimolar quantities by amino acid analysis, while no ammonia but two moles of alanine and one mole of an unknown amino acid were observed in a hydrolysate of dihydro-AM-toxin I-MeOH. The isolation of one mole of pyruvic acid from the acid hydrolysate of AM-toxin I indicated that AM-toxin I might contain a dehydroalanine residue.[350,351]

The acid hydrolysate of AM-toxin I showed four components in thin-layer chromatography. Two of them were ether-soluble, and one of the ether-soluble components was positive in the color test with 2,4-dinitrophenylhydrazine. Its phenylhydrazone was identified as that of pyruvic acid. Another ether-soluble component was suggested to be 2-hydroxy-3-methylbutanoic acid (Hmb) on the basis of PMR. Its structure, including absolute configuration, was determined by the comparison of a derivative of the hydroxy acid coupled with L-valine methyl ester with an authentic sample, H–L-Hmb–L(or D)-Val–OMe, by gas-liquid chromatography, to be L-2-hydroxy-3-methylbutanoic acid. The remaining two components were water-soluble and one of them was readily identified as L-alanine. The other was a new amino acid already designated as alternamic acid by Okuno's group.[340,342] From the mass spectrum of the methyl ester of the dinitrophenylated new amino acid, its structure was elucidated as 2-amino-5-(*p*-methoxyphenyl)pentanoic acid (Amp).[346] A study on oxidation of the amino acid with L-amino acid oxidase showed that the compound has L-configuration.

The amino acid sequence of AM-toxin I was elucidated from the mass spectra of AM-toxin I-MeOH (**8.1**) and dihydro-AM-toxin I-MeOH.[345,346] The spectra showed not only the sequence of the four components, but also provided definitive evidence that the lactone linkage is located between the hydroxy group of L-2-hydroxy-3-methylbutanoic acid and the carboxyl group of the L-alanine residue. The complete sequence of AM-toxin I is presented in Fig. 8.5.

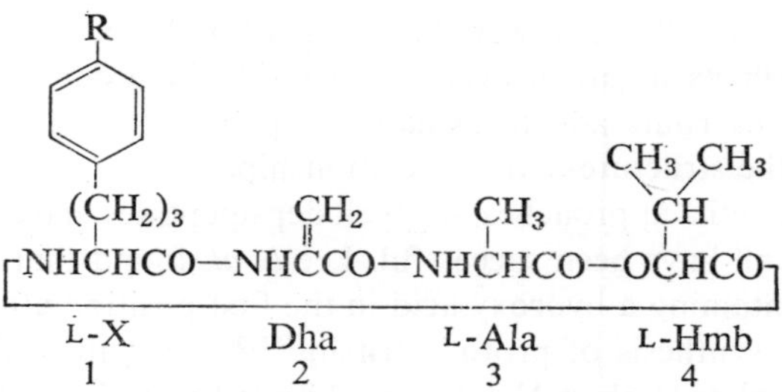

	R	X
AM-toxin I	CH_3O	Amp
AM-toxin II	H	App
AM-toxin III	HO	Ahp

Fig. 8.5 Primary structures of AM-toxins.

B. Primary Structures of AM-toxins II and III

By comparing the NMR, and mass spectra and the constituent amino acids of AM-toxin II with those of AM-toxin I, the amino acid sequence of AM-toxin II was concluded to be cyclo(–App–Dha–Ala–Hmb–) as shown in Fig. 8.5, where App is a 2-amino-5-phenylpentanoic acid residue.[346] Furthermore, Ueno *et al.* established the primary structure of AM-toxin III as cyclo(–Ahp–Dha–Ala–Hmb–), as shown in Fig. 8.5, where Ahp is 2-amino-5-(*p*-hydroxyphenyl)pentanoic acid, using AM-toxin III-MeOH derived from

AM-toxin III by methanolysis by the same procedure used in the case of AM-toxin I.[347] They suggested that the absolute configurations of the constituents in AM-toxins II and III might be the same as those of AM-toxin I without any definite evidence. Recently, they reported detailed mass spectrometric data for AM-toxins and related compounds.[352]

8.2 Chemical Synthesis of AM-toxins I, II and Their Analogs

8.2.1 PRELIMINARY STUDIES FOR THE SYNTHESIS OF AM-TOXINS

Organic syntheses of AM-toxins and related compounds have been carried out exclusively in the authors' laboratory. We intended to establish a synthetic confirmation of the structure of AM-toxins and to elucidate the structure-activity relationship. Prior to synthetic confirmation of the structure of AM-toxin I, because of difficulty in the synthesis of L-2-amino-5-(*p*-methoxyphenyl)pentanoic acid (L-Amp) and the comparative ease of preparation of *O*-methyl-L-tyrosine,[353] a lower homolog of L-Amp, we selected [L-Tyr(Me)^1]-AM-toxin as a model cyclodepsipeptide in order to search for an effective synthetic route and to exmaine its possible toxin activity from the standpoint of the structure-activity relationship.

Many synthetic approaches to cyclodepsipeptides have been attempted,[354] but not all have been successful. Lee *et al.* found that none of the intermediates containing a hydroxy acid in the first position afforded a desired product in the synthesis of protodestruxin.[355,356] Schulz also reported that H–D-Hmb–D-Val–D-Hmb–D-Val–OH could not be cyclized at the ester bond by the carbonyldiimidazole method, while cyclization of H–D-Val–D-Hmb–D-Val–D-Hmb–OH by the acid chloride method gave a cyclic dimer exclusively.[357] These findings indicated that the use of sequences with a hydroxy acid residue at the first position would be unfavorable.

Gross *et al.* succeeded in the preparation of Boc–L-Leu–L-Ala–Gly–Dha–$OCH_2C_6H_4$–resin from Boc–Gly–Dha–$OCH_2C_6H_4$–resin by the usual solid-phase method,[358] indicating that a dehydroalanine residue is stable to treatment with trifluoroacetic acid and triethylamine. Moreover, Rich and Mathiaparanam obtained tentoxin by cyclization from the corresponding tetrapeptide active ester containing an *N*-methyl-dehydrophenylalanine residue.[359] Therefore, a tetradepsipeptide active ester (**8.2**) was first subjected to

H–L-Tyr(Me)–Dha–L-Ala–L-Hmb–ONSu

(8.2)

cyclization in pyridine in the expectation that **8.2** might directly give the desired product. However, no cyclic monomer could be detected. The cyclization of two depsipeptide azides (**8.3** and **8.4**) also failed to afford a cyclic te-

H–L-Ala–L-Hmb–L-Tyr(Me)–Dha–N_3

(8.3)

H–L-Ala–L-Hmb–L-Tyr(Me)–L-Ser(Tos)–N_3

(8.4)

tradepsipeptide.[2,360,361] To account for the absence of a cyclic monomer product from **8.2–8.4,** we assumed that the orientation of the reactive side chain of dehydroalanine or the L-Ser(Tos) residue and/or the sequence of **8.2–8.4** might be unfavorable to the desired reaction.[361]

Photaki described the preparation of Z–Dha–Gly–OEt from Z–L-Ser (Tos)–Gly–OEt by the action of diethylamine in good yield,[362] suggesting that a cyclodepsipeptide with a serine residue might be converted into the corresponding peptide with a dehydroalanine residue. Hence, we prepared three peptide active esters (**8.5–8.7**),[2,360,361] because we had observed that the

H–L-Ser–L-Ala–L-Hmb–L-Tyr(Me)–ONSu

(8.5)

H–L-Tyr(Me)–L-Ser–L-Ala–L-Hmb–ONSu

(8.6)

H–L-Ala–L-Hmb–L-Tyr(Me)–L-Ser–ONSu

(8.7)

use of linear peptide succinimide esters gave high yield, simplified the experimental procedure, and suppressed racemization at the *C*-terminal residue in the preparation of many cyclopeptides related to gramicidin S.[68] The synthetic scheme for [L-Tyr(Me)[1]]-AM-toxin *via* **8.5** is illustrated in Fig. 8.6. Boc-tetrapeptide ester (**8.8**) was hydrogenated and converted to the active ester, then it was converted to the trifluoroacetate (**8.5**·TFA) of the linear peptide. Cyclization of **8.5**·TFA in pyridine provided a solid (**8.9**), but thin-layer chromatography revealed the presence of two major components in **8.9** as shown in Fig. 8.7. Fortunately the two components had fairly different characteristics on treatment with ethyl acetate, being separated into a less soluble component and a more soluble one. The less soluble component was purified by recrystallization and the purified compound was identified as a cyclic monomer by mass spectrometric determination of the molecular wiehgt.

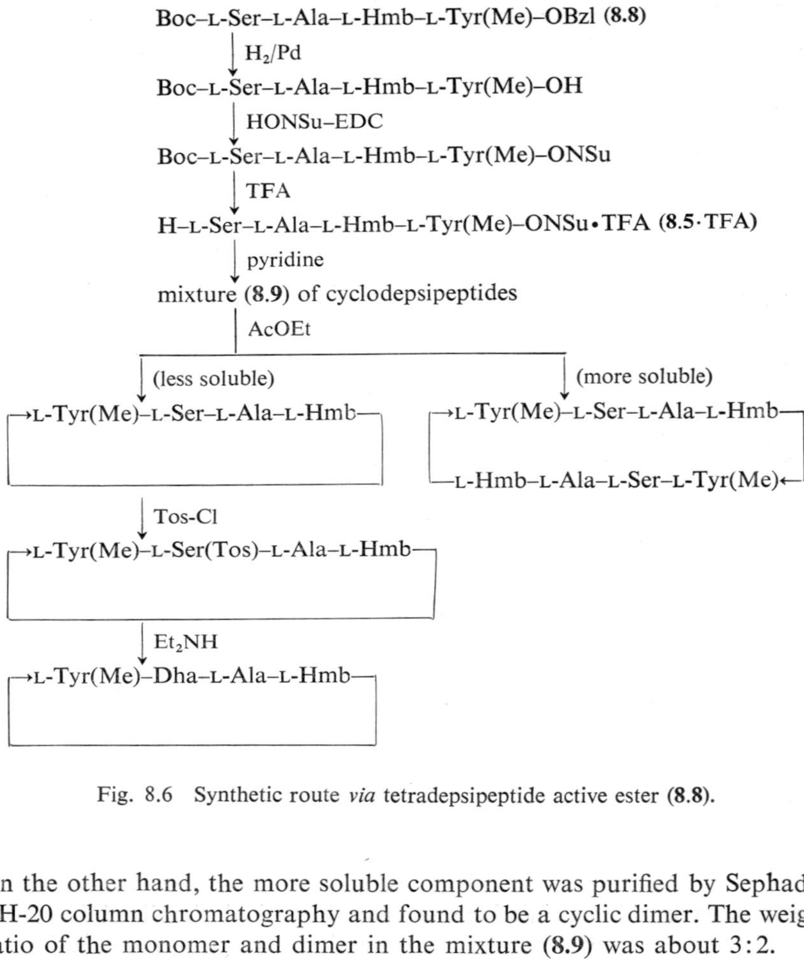

Fig. 8.6 Synthetic route *via* tetradepsipeptide active ester (**8.8**).

On the other hand, the more soluble component was purified by Sephadex LH-20 column chromatography and found to be a cyclic dimer. The weight ratio of the monomer and dimer in the mixture (**8.9**) was about 3:2.

The monomer was then converted into a cyclopeptide with a Ser(Tos) residue using tosyl chloride according to the method of Photaki.[362] When the Ser(Tos)-containing peptide was treated with diethylamine, the product exhibited many spots on a thin-layer chromatogram, including the desired compound, which was finally isolated by silica gel chromatography. The yield of [L-Tyr(Me)^{1}]-AM-toxin was calculated to be 16% from the cyclic peptide possessing a serine residue.[2,360,361]

We again applied the same procedure used for **8.5** to **8.6** and **8.7**;[361] however, we found that the product (**8.10** or **8.11**) from **8.6** or **8.7** formed predominantly cyclic polymers. The product (**8.10** or **8.11**) included a considerable amount of cyclic dimer and a small amount of cyclic monomer, as shown

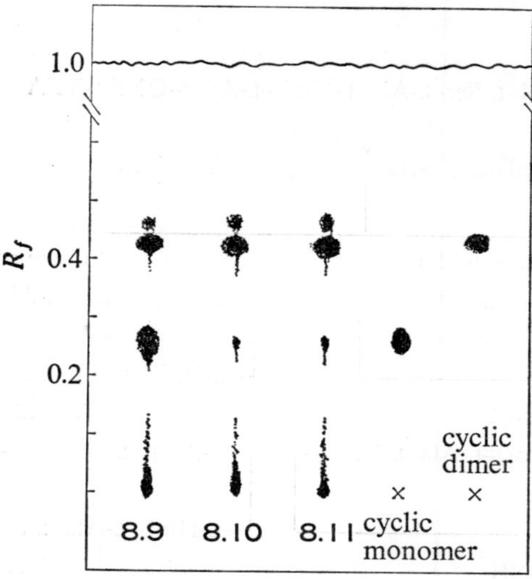

Fig. 8.7 Thin-layer chromatography of the crude products (**8.9**, **8.10**, and **8.11**) after cyclization of **8.5**, **8.6**, and **8.7** as well as the purified cyclodepsipeptides. Solvent, $CHCl_3$–MeOH (9:1). (Source: ref. 361. Reproduced by kind permission of Munksgaard)

in Fig. 8.7. These results are compatible with the observation in the synthesis of protodestruxin by Lee and Izumiya[356)] that the location of an ester bond far from either the *N*- or *C*-terminus was a significant factor in yielding a cyclodepsipeptide.

8.2.2 SYNTHESIS OF AM-TOXIN I AND ITS ANALOGS

A. Synthesis of AM-toxin I

On the basis of the findings described above, our group started the synthesis of a cyclodepsipeptide corresponding to the natural AM-toxin I.[2,360,363)] The synthetic scheme illustrated in Fig. 8.8 is quite similar to that for the synthesis of [L-Tyr(Me)^1]-AM-toxin. Synthesis of L-forms of 2-amino-5-(*p*-methoxyphenyl)pentanoic acid (Amp), 2-amino-5-phenylpentanoic acid (App) and 2-amino-5-(*p*-hydroxyphenyl)pentanoic acid (Ahp), which are constituents of AM-toxins I, II and III, respectively, was carried out as shown in Fig. 8.9.[364)] Acetyl-DL-Amp (**8.15a**) and acetyl-DL-App (**8.15b**) were resolved with acylase to afford their L-forms. L-Ahp was obtained from L-Amp by boiling in 6 M HCl at 110° C for 3 days. On the other hand, Okuno *et al.* reported the synthesis of DL-Amp by the Strecker method. Further, Ueno *et al.*[346)] and Okuno *et al.*[340)] isolated L-Amp from the acid hydrolysate of

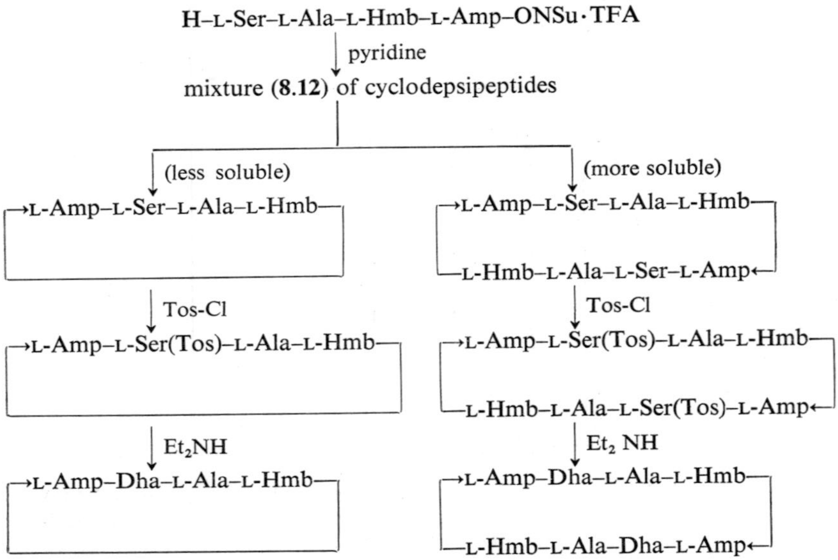

Fig. 8.8 Syntheses of AM-toxin I and its dimer *via* a linear tetradepsipeptide active ester.

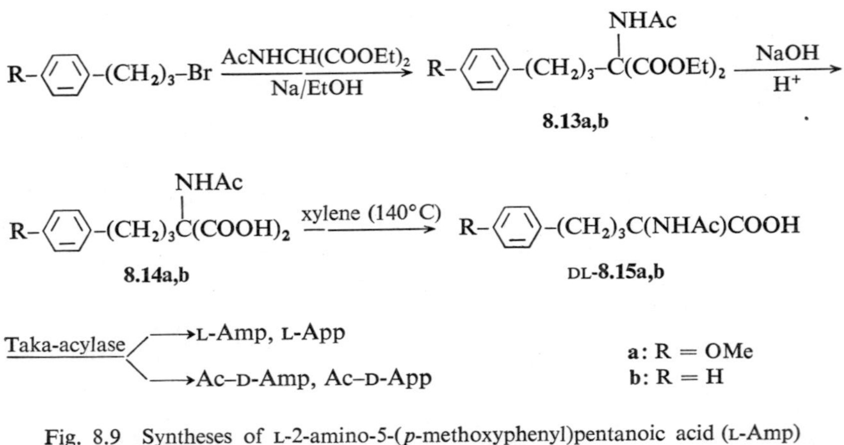

Fig. 8.9 Syntheses of L-2-amino-5-(*p*-methoxyphenyl)pentanoic acid (L-Amp) and L-2-amino-5-phenylpentanoic acid (L-App).

AM-toxin I, but they reported neither the occurrence nor isolation of Ahp from the acid hydrolysate.

The cyclized product (**8.12**) from H–Ser–Ala–Hmb–Amp–ONSu·TFA showed two major spots on a thin-layer chromatogram, as well as additional faint spots corresponding to by-products. After treatment with ethyl acetate, the less soluble component obtained was identified as a cyclic monomer (21% yield) with mp 193–195° C, $[\alpha]^{20}_{D}$ −113° (dimethylformamide). Using the same procedure described for the preparation of [L-Tyr(Me)^{1}]-AM-toxin, we obtained the desired cyclic peptide.[363] The synthetic compound had the same R_f value on a thin-layer chromatogram, and the same UV pattern (Fig. 8.10) and mass spectrum as natural AM-toxin I. The more soluble component was a cyclic dimer, and the weight ratio of monomer to dimer in the mixture (**8.12**) was found to be about 1 : 1. A dimerized form of AM-toxin I was also prepared from the dimer containing serine.

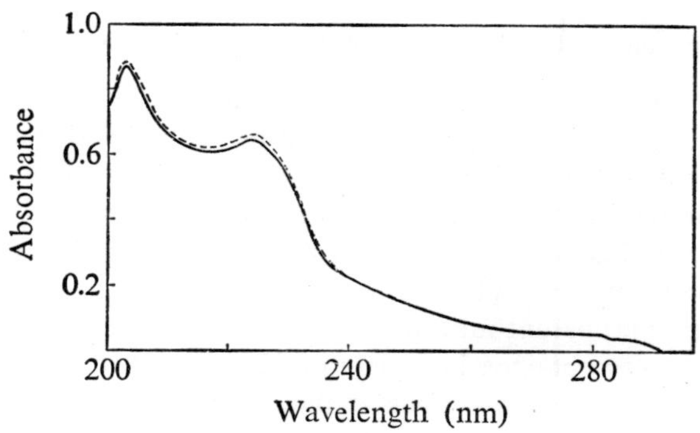

Fig. 8.10 UV patterns of synthetic (——) and natural (----) AM-toxin I. The concentration of each cyclopeptide in MeOH was 4.3×10^{-3} M. (Source: ref. 363. Reproduced by kind permission of Munksgaard)

B. Synthesis of AM-toxin Analogs

In order to clarify the structure-activity relationship of a biologically active peptide, it is often effective to synthesize analogs of the peptide and to examine their biological activity and conformation. Rich *et al.* showed that dihydrotentoxin had no activity to cause chlorosis, indicating that the double bond in tentoxin is essential for biological activity.[365] Therefore, Shimohigashi *et al.* were also interested in studying the role of the double bond in the dehydroalanine residue of AM-toxin I. Prior to the synthetic work, they attempted the catalytic hydrogenation of natural AM-toxin I and obtained a mixture of [L-Ala^{2}]- and [D-Ala^{2}]-AM-toxin I in a ratio of 4:6.[366,367]

Chemical synthesis of [L-Ala²]- and [D-Ala²]-AM-toxin I was carried out by a conventional method as described for the preparation of [L-Tyr(Me)¹]-AM-toxin. In addition, two other analogs containing *O*-methyl-L-tyrosine in place of L-2-amino-5-(*p*-methoxyphenyl)pentanoic acid, [L-Tyr(Me)¹, L-Ala²]- and [L-Tyr(Me)¹, D-Ala²]-AM-toxin, were also synthesized.[366,367] The results of activity tests will be described in Section 8.3.1. In the synthesis of [L-Ala²]-AM-toxin I, the cyclization of H–L-Ala–L-Ala–L-Hmb–L-Amp–ONSu•TFA afforded a mixture of cyclodepsipeptides. A thin-layer chromatogram of the mixture showed three spots corresponding to cyclic monomer, dimer and trimer. They found that dioxane was a good solvent for the separation of these three compounds. The less soluble fraction of the mixture in dioxane was further precipitated from dimethylformamide–dioxane to give the monomer (27 %) and trimer (8 %). Similar treatment of the soluble fraction gave the dimer (17 %). On the other hand, the cyclization of H–D-Ala–L-Ala–L-Hmb–L-Amp–ONSu surprisingly resulted in exclusive formation of the monomer (72 %). Compounds containing *O*-methyl-L-tyrosine were obtained similarly.[366,367]

C. Consideration of the Cyclization of Precursors Related to AM-toxin

The fruitful results obtained through the syntheses of AM-toxin I and its analogs cast light not only on conformation of peptide active esters during reaction but were also helpful in developing a strategy for further syntheses of AM-toxins and related peptides. In the case of the synthesis of [L-Tyr(Me)¹]-AM-toxin, cyclization *via* the Ser–Ala or Hmb–Tyr(Me) bond, which is next to the ester bond, gave only a trace amount of monomer and a large amount of cyclic polymer (apparent ratio of monomer to polymer, 1:20). However, cyclic monomer was obtained in a yield of 32 % and cyclic dimer in a yield 11 % when the cyclization was done at the Tyr(Me)–Ser bond.

Kato *et al.* suggested that as the constituent amino acids have no *β*-branched side chains, the difference in cyclization yield could not be attributed to steric hindrance of the reactants.[368] The linear tetrapeptide with the L-Ser–L-Ala–L-Hmb–L-Tyr(Me) sequence was supposed to have a favorable conformation to yield cyclic monomer. However, the ORD patterns of the linear peptide esters were found to be similar to each other, but completely different from those of AM-toxin I (a trough at 250 nm, $[\alpha] = -6{,}000°$) and [L-Tyr-(Me)¹]-AM-toxin (242 nm, $[\alpha] = -11{,}000°$). Next, Kato *et al.*[368] found that the difference in cyclization yield could be well explained by assuming the Ser–Ala and Hmb–Tyr(Me) bonds to be *cis* and the Tyr(Me)–Ser bond to be *trans*. Cyclization *via* the formation of an unfavorable *cis* bond was difficult and the reaction proceeded to give polymers. In the case of *trans* bond formation, cyclization gave a reasonable amount of cyclic monomer and dimer without difficulty in a concentration-dependent ratio. The remaining Ala–Hmb ester bond should be *trans*, because a *trans* ester bond was shown to be much more stable than a *cis* bond. This may also be applicable in the case of all-*trans* backbone conformation.

On the other hand, the results obtained by Shimohigashi *et al.*[366,367] in the synthesis of a D- or L-Ala^2-containing cyclodepsipeptide indicate that not only the sequence of the linear peptides but also the presence of a D-amino acid residue at the *N*-terminus is important to obtain cyclic monomer in high yield. In particular, the exclusive formation of monomer together with NMR data suggests that the presence of a D-amino acid makes the molecule take a conformation favorable for the formation of a reaction intermediate suitable for ready monomerization.

8.2.3 SYNTHESIS OF AM-TOXIN II AND ITS ANALOGS

Our group reported the synthesis of AM-toxin II and its analog, [L-Phe^1]-AM-toxin, recently.[369] The synthetic route to AM-toxin II is illustrated in Fig. 8.11. As it had already been found that the presence of a hydroxy acid residue in the third position from the *N*-terminus and a D-amino acid residue at the *N*-terminus of the linear precursor favored the cyclization reaction, we selected the sequence H–D-Ser–L-Ala–L-Hmb(–L-App–ONSu) as a linear despipeptide precursor. As expected, cyclic monomer was obtained in

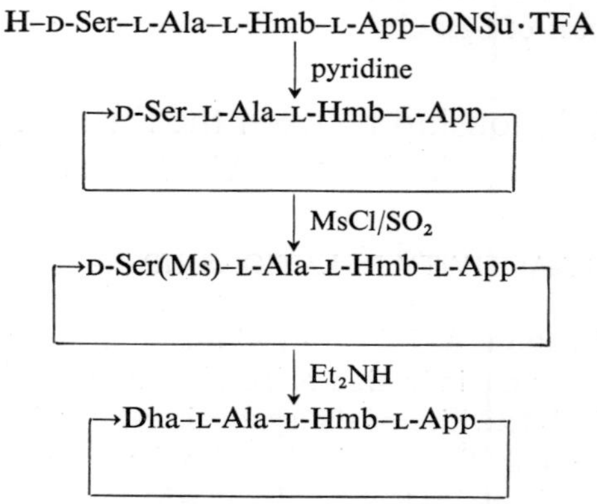

Fig. 8.11 Synthesis of AM-toxin II.

good yield. However, at the step of conversion of a serine residue into a dehydroalanine residue, it was observed that the tosylation of cyclodepsipeptide containing a serine residue was difficult, and the tosylated product was quite stable to treatment with bases such as diethylamine or piperazine. The use of excess base or boiling in pyridine[370] resulted in decomposition of the tosylated compound. Phenylmethanesulfonyl chloride **(8.16)**[371] or methyl chlorosulfite (**8.17**)[372] hardly reacted with cyclo(–L-App–D-Ser–L-Ala–L-Hmb–). When methanesulfonyl chloride containing sulfur dioxide (MsCl/

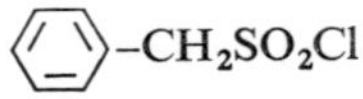

(**8.16**)

CH_3OSOCl

(**8.17**)

SO_2)[373] was used, the methanesulfonylated compound, [D-Ser(Ms)^2]-AM-toxin II was obtained in moderate yield. Sulfur dioxide might act as a catalyst, because methanesulfonyl chloride itself did not react with cyclo (–L-App–D-Ser–L-Ala–L-Hmb–).

[D-Ser(Ms)^2]-AM-toxin II was converted to the desired AM-toxin II by treatment with diethylamine, and [L-Phe^1]-AM-toxin was obtained similarly. Synthetic AM-toxin II with mp 212–213° C (reported value, 213–214° C[346]) showed a molecular weight identical with the calculated value on mass spectrometry. In addition to having the same R_f in thin-layer chromatography, synthetic and natural AM-toxin II showed the same mass spectra, UV pattern and crystal form (needles). From these results, we proposed that the difference in sulfonic ester-forming behavior between cyclodepsipeptides containing D-serine and L-serine may be attributed to a difference in their conformation,[366] and that the difficulty in the elimination of tosyl and phenylmethanesulfonyl groups may be due to steric hindrance of their bulky phenyl groups preventing the access of bases.[369]

8.3 Relationship between the Structure and Activity of AM-toxins

8.3.1 STRUCTURE-ACTIVITY RELATIONSHIP

Synthetic and natural AM-toxin I had the same minimum toxic activity of 2×10^{-3} μg/ml for the induction of necrosis on an apple leaf. Table 8.1 shows the biological activities of the synthetic AM-toxins on apple leaves. From Table 8.1, not only the ring size but also the presence of the bulky and hydrophobic side chain of the Amp residue in AM-toxin I seems to be important for the exhibition of biological activity; in other words, the interaction between AM-toxin I and a possible receptor in an apple leaf is very specific.

The minimum toxic activity (2×10^{-2} μg/ml) of synthetic AM-toxin II was identical with that of the natural product.[369] Catalytic hydrogenation of the double bond in the dehydroalanine residue of AM-toxins resulted in a lowering the activity about 20,000–50,000 times,[366,367] indicating that the presence of the double bond in dehydroalanine is of importance. Rich *et al.* also reported that tentoxin, a phytotoxic cyclotetrapeptide, lost its activity

TABLE 8.1 Synthetic AM-toxins and Their Activities on Apple Leaves

AM-toxin	Side chain: Position 1	Side chain: Position 2	Necrotic activity ($\mu g/ml$)
AM-toxin I	$-CH_2-CH_2-CH_2-C_6H_4-OCH_3$	$=CH_2$	0.002
AM-toxin II	$-CH_2-CH_2-CH_2-C_6H_5$	$=CH_2$	0.02
[D-Ala^2]-AM-toxin I	$-CH_2-CH_2-CH_2-C_6H_4-OCH_3$	$-CH_3$ (D)	5–10
[L-Phe^1]-AM-toxin I	$-CH_2-C_6H_5$	$=CH_2$	20
AM-toxin I dimer	$(-CH_2-CH_2-CH_2-C_6H_4-OCH_3)$	$(=CH_2)$	20–40
[L-Tyr(Me)^1]-AM-toxin	$-CH_2-C_6H_4-OCH_3$	$=CH_2$	20–50
[L-Ser^2]-AM-toxin I	$-CH_2-CH_2-CH_2-C_6H_4-OCH_3$	$-CH_2OH$ (L)	30
[L-Tyr(Me)^1, D-Ala^2]-AM-toxin	$-CH_2-C_6H_4-OCH_3$	$-CH_3$ (D)	>100
[L-Ala^2]-AM-toxin I	$-CH_2-CH_2-CH_2-C_6H_4-OCH_3$	$-CH_3$ (L)	>100
[L-Tyr(Me)^1, L-Ala^2]-AM-toxin	$-CH_2-C_6H_4-OCH_3$	$-CH_3$ (L)	>100

upon hydrogenation, even at high concentrations.[365] Table 8.1 shows that [D-Ala^2]-AM-toxin I retained very low but recognizable activity, whereas [L-Ala^2]-AM-toxin I no longer showed activity.[366,367] This shows that the presence of a double bond is not the only factor affecting the activity.

The NMR study of [D-Ala^2]-AM-toxin I and [L-Tyr(Me)^1, D-Ala^2]-AM-toxin by Shimohigashi *et al.*[366,367] indicated the presence of a common conformer. However, the content of the same conformer of [L-Ala^2]-AM-toxin I and [L-Tyr(Me)^1, L-Ala^2]-AM-toxin, if any, must be extremely low. This observation suggests that there is a specific conformation related to biological activity. We will next describe a study on the conformations of the analogs.

8.3.2 CONFORMATIONS OF AM-TOXIN AND ITS ANALOGS

Ueno *et al.*[346] observed two α-protons as unresolved multiplets at 4.3 ppm for AM-toxins I and II in DMSO-d_6 and two protons as doublets at 7.9 and 8.0 ppm, for which coupling constants of 8 and 9 Hz were given by Okuno *et al.*[342] The chemical shifts of α-protons (4.3 ppm) were commonly observed in peptide systems, and no low field signals could be observed around the 5 ppm region.

As described already, the analog containing a D-alanine residue in place of a dehydroalanine residue in AM-toxin I shows weak but recognizable toxicity, while the analog containing L-alanine does not. As the difference in toxicity indicated that a definite conformation is important for the activity, Shimohigashi *et al.*[366,367] carried out conformational analysis of AM-toxin I and related compounds, mainly using NMR. The spectra of [D-Ala^2]-AM-toxin I and [L-Tyr(Me)^1, D-Ala^2]-AM-toxin in trifluoroacetic acid showed well-resolved and similar patterns, indicating the presence of a stable con-

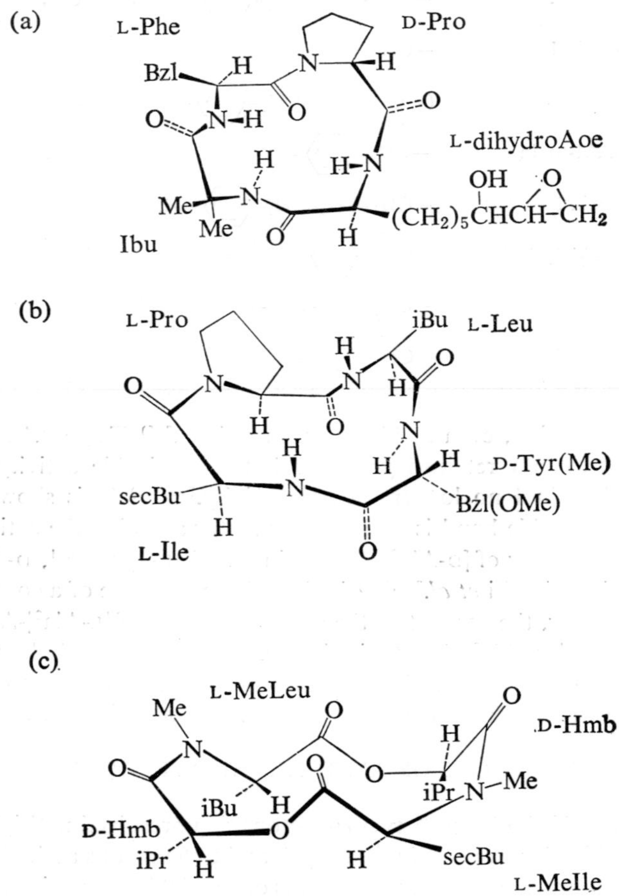

Fig. 8.12 Conformations of cyclotetrapeptides. (Source: ref. 374. Reproduced by kind permission of John Wiley & Sons) (Source: ref. 375. Reproduced by kind permission of The Chemical Society of Japan) (Source: ref. 376. Reproduced by kind permission of The American Chemical Society)

former for both peptides as well as similarity of the conformations. Furthermore, characteristic features such as coupling constants and chemical shifts of NH and α-CH protons are similar to those of AM-toxin I. These physicochemical and biological findings suggest that the common conformation of the two toxic analogs is similar to the conformation of AM-toxins.

So far, a variety of backbone conformations have been proposed for natural and synthetic cyclotetrapeptides. Among them, at least three conformations are consistent with the results obtained by NMR studies: (1) conformation a with an all-*trans* backbone system; (2) conformation b with a *trans-trans-trans-cis* system and (3) conformation c with a *cis-trans-cis-trans* backbone system. Conformation a was proposed for dihydrochlamydocin,[374)] conformation b for an analog of Cyl-2,[375)] and conformation c for a synthetic depsipeptide, cyclo(–D-Hmb–L-MeIle–D-Hmb–L-MeLeu–),[376)] as shown in Fig. 8.12. Generally, the presence of a *cis* bond is likely for a bond containing an *N*-alkyl amino acid such as proline. Therefore, an all-*trans* backbone seems to be most probable. In the case of a cyclic tetrapeptide, however, an all-*trans* backbone will be strained because of steric constraint.[377)] The presence of conformation b or c containing a *cis* bond thus cannot be excluded.

A possible backbone conformation of [D-Ala^2]-AM-toxin I with backbone a is illustrated in Fig. 8.13.[378)] In order to obtain further information, new techniques such as NOE (Section 5.4.3) should be useful. Further studies are in progress in Izumiya's laboratory.

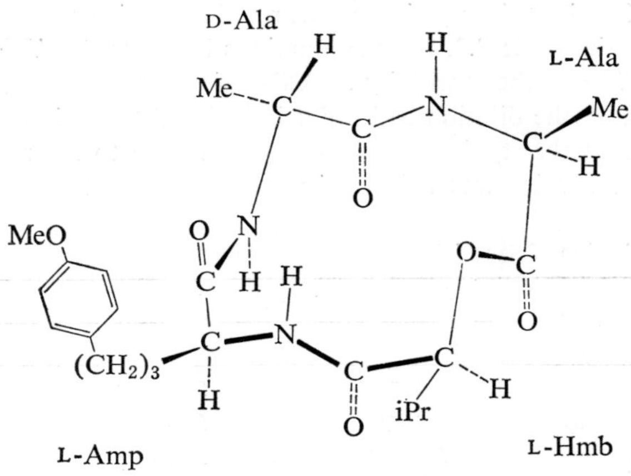

Fig. 8.13 Possible backbone conformation of [D-Ala^2]-AM-toxin I.

Inactive analogs containing L-alanine shows different features. The NMR spectra of [L-Ala^2]-AM-toxin I and [L-Tyr(Me)^1, L-Ala^2]-AM-toxin in trifluoroacetic acid showed complex patterns, which indicated that these peptides were mixtures of more than two conformers. We concluded that the main portions of the analogs containing L-Ala^2 have different conformations from those of [D-Ala^2]-AM-toxin I and [L-Tyr(Me)^1, D-Ala^2]-AM-toxin, based on the extremely low intensities of the signals which have nearly the same chemical shifts as those of [D-Ala^2]-AM-toxin I and [L-Tyr(Me)^1, D-Ala^2]-AM-toxin.[366,367]

The ORD study by Shimohigashi *et al.*[366,367] showed a significant trough, $[\alpha] = -6000$ to $-9000°$, around the 245 nm region for all the above cyclic tetradepsipeptides. In contrast, dimers of [L-Ala^2]-AM-toxin I and [L-Tyr (Me)^1, L-Ala^2]-AM-toxin gave a shallow trough, less than $-2500°$, at around 245 nm. These results suggest that the peptide chromophores in cyclic tetradepsipeptides have a certain distinct spatial orientation, but chromophores in peptides with larger ring structures do not.

8.4 Biological Study of AM-toxins

When an aqueous solution of AM-toxin I was dropped on an apple leaf, the formation of necrosis was observed and the symptoms were the same as those caused by inoculation of spores of *Alternaria mali*. Only plants susceptible to the toxicity of AM-toxin I were infected by a culture filtrate from *Alternaria mali* (strain I-716), which is very virulent. These observations mean that AM-toxin I plays an important role in selective recognition in the host. As shown in Table 8.2, the examinations by Ueno *et al.*[345,379,380] and Kohmoto *et al.*[381] of the sensitivity to AM-toxin among several cultivars of apple showed that the dilution end point inducing Alternaria blotch varied continuously from the sensitive cultivars such as Indo to the resistant ones such as Jonathan, which was 10^4–10^5 times more resistant than Indo.[345,379]

TABLE 8.2 Corrleation of Sensitivities to AM-toxins in Several Cultivars of Apple

Cultivar	AM-toxin (μg/ml)		
	I	II	III
Indo	10^{-5}	5×10^{-3}	10^{-4}
Starking Delicious	10^{-3}	5×10^{-2}	
Fuji	5×10^{-3}	5×10^{-2}	
Kokko	5×10^{-2}	5×10^{-2}	
Golden Delicious	10^{-2}	5×10^{-2}	
Iwai	10^{-1}	5×10^{-1}	
Asahi	1	5×10^{-1}	
Jonathan	1	>1	10

The continuous distribution of the toxin sensitivity is compatible with the suggestion that the susceptibility of the apple to *Alternaria mali* may be conditioned by multiple genes.[382,383]

Kohmoto *et al.* recently examined the production and release of host-specific toxins by spores of *Alternaria mali*.[348] When water extracts from resting spores were tested for host-specific toxicity using the apple leaf bioassay, none of the extracts was toxic, indicating that resting spores did not contain any free or weakly-bound host-specific toxins that could be released easily into water. To examine the release of host-specific toxins from germinating spores, a spore suspension was incubated at 28° C and the percentage germination of the spores was determined at various times. It was found that almost 90% of the spores had germinated by 3 h, and the toxins were present. Moreover, they observed that the toxicity increased rapidly during 2–6 h in the incubation period, and thin-layer chromatography of the toxins revealed that AM-toxins I and II were major products in addition to several minor toxins.[348]

In order to clarify the role of AM-toxins in the course of infection of a host plant, Sekiguchi *et al.*[384] inoculated spores of *Alternaria mali* into young leaves from the susceptible apple cv. Indo and observed the morphological changes of the leaves. After 2–3 h, spore germination started and then the tissues contacting the hyphae, which produce the toxins, became dark, and 12–24 h later the infecting hyphae penetrated the epidermal cells of the susceptible leaves of the host plant. These observations suggested that proir to intrusion of the hyphae, AM-toxins might be secreted to blotch the tissues.[384]

Kohmoto *et al.*[348] demonstrated that AM-toxin I at 10^{-5} M caused an increase in the loss of electrolytes from susceptible leaves. A dose-response study showed that AM-toxin I induced an increase of electrolyte loss even at approximately 10^{-8} M in susceptible tissues, and at concentrations of over 10^{-5} M in resistant ones. However, AM-toxin I did not have any appreciable effect on respiration. Other AM-toxins also caused an increase in electrolyte loss from leaves of susceptible cultivars.[348] When *Alternaria mali* was used instead of the toxins a small but significant increase in electrolyte loss was observed only from susceptible tissues at 3–4 h after inoculation.

The same researchers observed in a microscopical examination that necrotic browning of the susceptible leaves occurred on the epidermal cells beneath appressoria, and also on the neighboring cells with the advance of the hyphal invasion at 9–10 h after inoculation. At about the same time, the leaves showed a clear increase of electrolyte loss. These results imply that prior to host invasion, AM-toxins exert an initial effect on the plasma membrane, leading to a slight disturbance in the host cells.[348] This is consistent with the hypothesis that AK-toxin may play a role as the initial inciting agent in black spot of Japanese pear.[385,386]

CHAPTER **9**

Concluding Remarks

Throughout this book, we have emphasized that chemical synthesis of analogs is a useful method for the study of structure-activity relationships of biologically active peptides, e.g., gramicidin S and AM-toxins. Such a study can give wider and deeper insight, which could not be obtained by the mere use of naturally occurring peptides and their modifications for an understanding of the action of biologically active peptides.

After the first synthesis of gramicidin S by Schwyzer and Sieber, nearly a hundred related compounds, including synthetic analogs and chemically modified derivatives of natural gramicidin S were prepared. The results of studies with these preparations on structure-activity relationships show that important factors for antimicrobial activity are a hydrophilic-hydrophobic "sided" structure, an appropriate distance between two basic groups, and rigid β-turns which stabilize the antiparallel β-pleated sheet structure. These characteristic features allow the peptides to bind with the negatively charged bilayers of sensitive microorganisms by means of electrostatic and hydrophobic interactions, and finally to destroy the bilayer. On the basis of these findings, we can conclude that structure-activity relationships of gramicidin S are relatively well understood, although the molecular mechanism of membrane cleavage remains to be elucidated. We will now report the results of recent studies and discuss the remaining problems.

At the Fifth American Peptide Symposium, Ivanov suggested that gramicidin S interacts with the lipid bilayer to form a hydrophobic complex. In the complex, gramicidin S is assumed to hold an α,β-mixed conformation.[229] It would be interesting to obtain experimental evidence of this conformation, because the possible occurrence of α,β-structure was discussed by Balasubramanian[236] in 1967, but no evidence has yet been found for its occurrence. Such a finding should help to clarify the mode of action of this antibiotic *in vivo*.

Hull *et al.*[387] studied the conformation of crystals of hydrated gramicidin S obtained from aqueous urea and reported that the crystals had a folded β-sheet conformation. Interestingly, this result appears to be compatible with the results of recent studies on protein crystals, which show that antiparallel β-sheet structures in several proteins are folded right-handedly.[388]

Shemyakin and Ovchinnikov first proposed the concept of retroenantiomeric peptides (Section 5.8.2). Ovchinnikov later showed that the antimicrobial activities of enantiomers of gramicidin S and [$Gly^{5,5'}$]-gramicidin S were identical to, and the activities of retrogramicidin S and retro-[$Gly^{5,5'}$]-gramicidin S were much lower than, those of gramicidin S and [$Gly^{5,5'}$]-gramicidin S, respectively (Table 5.2). Miroshnikov *et al.*[145] studied the conformations of topological analogs of gramicidin S and concluded that the conformations of gramicidin S and [$Gly^{5,5'}$]-gramicidin S differ from those of retrogramicidin S and retro-[$Gly^{5,5'}$]-gramicidin S. The results of CD, IR, and ^{1}H-NMR studies showed that the content of β-sheet structure in the retro-analogs, if any, should be much lower than that of the parent peptides. Since the conformation of enantiomeric analogs should be the same as that of the parent peptides, low activity of retro-analogs can be explained in terms of conformational differences. It is again clear that similarity in conformation is the decisive factor for activity.

Apart from gramicidin S itself, replacement analogs present their own problems. For example, in contrast to the fully active [L-$Leu^{1,1'}$]-gramicidin S,[107] [L-$Val^{3,3'}$]-gramicidin S shows only weak activity.[124] There is a distinct difference between substitutions at position 1 and position 3. The depth of the trough of [L-$Val^{3,3'}$]-gramicidin S at 233 mnm in its ORD curve was one-half of that of gramicidin S. A similar results was also obtained in the case of [L-$Phe^{1,1',3,3'}$]-gramicidin S.[127] These results suggest that the leucine residue, the third amino acid in a β-sheet region or the first position in a β-turn region, plays a more important role in stabilizing the conformation of gramicidin S than the valine residue. At present, we cannot explain this difference in activity between these 1,1′ and 3,3′ substitutions. Such a problem is indeed unique to the synthetic analogs, but if these new problems can be successfully solved, a better understanding of the mode of action of gramicidin S must surely be obtained.

Although some unresolved problems remain, it seems clear that synthetic studies on gramicidin S and its analogs have made a fruitful contribution to the understanding of structure-activity relationships of biologically active peptides. In addition, newly developed techiques of chemical and biochemical research will surely extend our understanding of the mechanism of gramicidin S action, which again will be significant in a much wider context.

References

1. T. Kato and N. Izumiya, *Chemistry and Biochemistry of Amino Acids, Peptides and Proteins* (ed. B. Weinstein), vol. 2, p. 1, Marcel Dekker Inc. (1974).
2. N. Izumiya, *Biologically Active Peptides Produced by Microorganisms* (in Japanese) (ed. H. Umezawa, T. Shiba and T. Takita), p. 166, Kyoritsu Shuppan (1976).
3. G. F. Gause and M. G. Brazhnikova, *Am. Rev. Sov. Med.*, **2**, 134 (1944); *Chem. Abstr.*, **39**, 1893 (1945).
4. G. F. Gause and M. G. Brazhnikova, *Lancet*, **247**, 715 (1944).
5. G. F. Gause and M. G. Brazhnikova, *Nature*, **154**, 703 (1944).
6. R. J. Dubos, *Proc. Soc. Exp. Biol. Med.*, **40**, 311 (1939).
7. R. J. Dubos, *J. Exp. Med.*, **70**, 1 (1939).
8. R. J. Dubos, *ibid.*, **70**, 11 (1939).
9. R. J. Dubos and C. Cattaneo, *ibid.*, **70**, 249 (1939).
10. R. D. Hotchkiss and R. J. Dubos, *J. Biol. Chem.*, **132**, 791 (1940).
11. R. D. Hotchkiss and R. J. Dubos, *ibid.*, **132**, 793 (1940).
12. R. D. Hotchkiss and R. J. Dubos, *ibid.*, **136**, 803 (1940).
13. M. A. Ruttenberg, T. P. King and L. C. Craig, *J. Am. Chem. Soc.*, **87**, 4196 (1965).
14. S. Otani, H. Nagano and Y. Saito, *J. Osaka City Med. Cent.* (in Japanese), **7**, 640 (1958).
15. A. R. Battersby and L. C. Craig, *J. Am. Chem. Soc.*, **73**, 1887 (1951).
16. R. Schwyzer and P. Sieber, *Helv. Chim. Acta*, **40**, 624 (1957).
17. M. Waki and N. Izumiya, *J. Am. Chem. Soc.*, **89**, 1278 (1967).
18. M. Waki and N. Izumiya, *Bull. Chem. Soc. Jpn.*, **40**, 1687 (1967).
19. R. L. M. Synge, *Biochem. J.*, **39**, 363 (1945).
20. R. L. M. Synge, *ibid.*, **65**, 750 (1957).
21. R. Consden, A. H. Gordon, A. J. P. Martin and R. L. M. Synge, *ibid.*, **41**, 596 (1947).
22. F. Sanger, *ibid.*, **40**, 261 (1946).
23. A. J. P. Martin and R. Mittelman, *ibid.*, **43**, 353 (1948).
24. K. O. Pedersen and R. L. M. Synge, *Acta Chem. Scand.*, **2**, 408 (1948).
25. A. N. Belozerskii and T. S. Paskhina, *Biokhimiya*, **10**, 344 (1945); *Chem. Abstr.*, **40**, 1557 (1946).
26. G. M. J. Schmidt, D. C. Hodgkin and B. M. Oughton, *Biochem. J.*, **65**, 744 (1957).
27. A. R. Battersby and L. C. Craig, *J. Am. Chem. Soc.*, **74**, 4019 (1952).
28. A. R. Battersby and L. C. Craig, *ibid.*, **74**, 4023 (1952).
29. A. Paladini and L. C. Craig, *ibid.*, **76**, 688 (1954).

30. T. P. King and L. C. Craig, *ibid.*, **77**, 6627 (1955).
31. M. A. Ruttenberg, T. P. King and L. C. Craig, *Biochemistry*, **4**, 11 (1965).
32. M. A. Ruttenberg and B. Mach, *ibid.*, **5**, 2864 (1966).
33. K. Fujikawa, T. Sakamoto, T. Suzuki and K. Kurahashi, *Biochim. Biophys. Acta*, **169**, 520 (1968).
34. R. D. Hotchkiss and R. J. Dubos, *J. Biol. Chem.*, **141**, 155 (1941).
35. R. D. Hotchkiss, *ibid.*, **141**, 171 (1941).
36. R. L. M. Synge, *Biochem. J.*, **39**, 355 (1945).
37. R. L. M. Synge, *Cold Spring Harbor Symp. Quant. Biol.*, **14**, 191 (1950).
38. J. D. Gregory and L. C. Craig, *J. Biol. Chem.*, **172**, 839 (1948).
39. L. C. Craig, J. D. Gregory and G. T. Barry, *Cold Spring Harbor Symp. Quant. Biol.*, **14**, 24 (1949).
40. L. K. Ramachandran, *Biochemistry*, **2**, 1138 (1963).
41. S. Ishii and B. Witkop, *J. Am. Chem. Soc.*, **85**, 1832 (1963).
42. S. Ishii and B. Witkop, *ibid.*, **86**, 1848 (1964).
43. R. Sarges and B. Witkop, *ibid.*, **86**, 1861 (1964).
44. R. Sarges and B. Witkop, *ibid.*, **86**, 1862 (1964).
45. R. Sarges and B. Witkop, *ibid.*, **87**, 2011 (1965).
46. R. Sarges and B. Witkop, *ibid.*, **87**, 2020 (1965).
47. R. Sarges and B. Witkop, *ibid.*, **87**, 2027 (1965).
48. R. Sarges and B. Witkop, *Biochemistry*, **4**, 2491 (1965).
49. E. Gross and B. Witkop, *ibid.*, **4**, 2495 (1965).
50. D. W. Urry, M. C. Goodall, J. D. Glickson and D. F. Mayers, *Proc. Natl. Acad. Sci. U.S.A.*, **68**, 1907 (1971).
51. D. W. Urry, J. D. Glickson, D. F. Mayers and J. Haider, *Biochemistry*, **11**, 487 (1972).
52. Y. Noda, *J. Chem. Soc. Jpn. Pure Chem. Sect.* (in Japanese), **79**, 663 (1958).
53. T. Kato and N. Izumiya, *J. Biochem.* (*Tokyo*), **59**, 629 (1966).
54. T. Kato and N. Izumiya, *Bull. Chem. Soc. Jpn.*, **39**, 2242 (1966).
55. K. Kurahashi, *J. Biochem.* (*Tokyo*), **56**, 101 (1964).
56. S. Otani and Y. Saito, *ibid.*, **56**, 103 (1964).
57. G. G. Zharikova, S. P. Myaskovskaya and A. B. Silaev, *Vestn. Mosk. Univ., Biol., Pochvoved.*, **27**, 110 (1972).
58. S. P. Myaskovskaya, G. G. Zharikova and A. B. Silaev, *ibid.*, **28**, 123 (1973).
59. V. du Vigneaud, C. Ressler, J. M. Swan, C. W. Roberts, P. G. Katsoyannis and S. Gordon, *J. Am. Chem. Soc.*, **75**, 4879 (1953).
60. R. Boissonnas and I. Schumann, *Helv. Chim. Acta*, **35**, 2229 (1952).
61. T. Kato, M. Ohno, S. Makisumi and N. Izumiya, *J. Chem. Soc. Jpn. Pure Chem. Sect.* (in Japanese), **87**, 493 (1966).
62. K. D. Kopple, *J. Pharm. Sci.*, **61**, 1345 (1972).
63. M. Rothe, R. Theysohn, D. Mühlhausen, F. Eisenbeiss and W. Schindler, *Chemistry and Biology of Peptides* (ed. J. Meienhofer), p. 51, Ann Arbor Science Pub. Inc. (1972).
64. R. Schwyzer and P. Sieber, *Helv. Chim. Acta*, **41**, 2186 (1958).
65. R. Schwyzer, *Amino Acids and Peptides with Antimetabolic Activity* (CIBA Found. Symp.), p. 171, Churchill (1958).

66. L. Pauling and R. B. Corey, *Proc. Natl. Acad. Sci. U.S.A.*, **39**, 247 (1953).
67. H. Aoyagi, M. Kondo, T. Kato, S. Makisumi and N. Izumiya, *Bull. Chem. Soc. Jpn.*, **40**, 1685 (1967).
68. S. Matsuura, H. Takiguchi, M. Waki and N. Izumiya, *Mem. Fac. Sci. Kyushu Univ., Ser. C*, **9**, 277 (1975).
69. M. Ohno, S. Tsukamoto, S. Makisumi and N. Izumiya, *Bull. Chem. Soc. Jpn.*, **45**, 2852 (1972).
70. K. Tanaka, A. Shichiho and S. Sakakibara, *Ann. Mtg. Chem. Soc. Japan* (in Japanese), 1967, *Abstracts*, III, p. 661.
71. K. Nonaka, K. Kamekawa, K. Sato, M. Waki, T. Kato and N. Izumiya, *Peptide Chemistry* (ed. T. Shiba), p. 135, Protein Research Foundation (1978).
72. H. Klostermeyer, *Chem. Ber.*, **101**, 2823 (1968).
73. G. Losse and K. Neubert, *Tetrahedron Lett.*, **1970**, 1267.
74. M. Ohno, K. Kuromizu, H. Ogawa and N. Izumiya, *J. Am. Chem. Soc.*, **93**, 5251 (1971).
75. F. Weygand and U. Ragnarsson, *Z. Naturforsch.*, **21b**, 1141 (1966).
76. K. Sato, H. Abe, T. Kato and N. Izumiya, *Bull. Chem. Soc. Jpn.*, **50**, 1999 (1977).
77. H. Aoyagi, T. Kato, M. Ohno and N. Izumiya, *J. Am. Chem. Soc.*, **86**, 5700 (1964).
78. H. Aoyagi, T. Kato, M. Ohno, M. Kondo, M. Waki, S. Makisumi and N. Izumiya, *Bull. Chem. Soc. Jpn.*, **38**, 2139 (1965).
79. K. Hofmann and H. Yajima, *J. Am. Chem. Soc.*, **83**, 2289 (1961).
80. M. Wilchek, S. Sarid and A. Patchornik, *Biochim. Biophys. Acta*, **104**, 616 (1965).
81. W. F. Benisek, M. A. Raftery and R. D. Cole, *Biochemistry*, **6**, 3780 (1967).
82. M. Kawai and U. Nagai, *Biopolymers*, **17**, 1549 (1978).
83. Yu. A. Ovchinnikov, A. A. Kiryushkin and I. V. Kozhevnikova, *Zh. Obshch. Khim.*, **38**, 2636 (1968).
84. M. Iwai, K. Nakajima, A. Uno, S. Hase, I. Takeuchi and K. Okawa, *Bull. Chem. Soc. Jpn.*, **43**, 3246 (1970).
85. N. A. Krit, A. L. Zhuze, T. M. Andronova, M. P. Filatova, G. A. Ravdel and L. A. Shchukina, *Bioorg. Khim.*, **1**, 247 (1975).
86. M. A. El-Maghraby, *J. Indian Chem. Soc.*, **53**, 496 (1976).
87. M. A. El-Maghraby, *Indlan J. Chem. Sect. B*, **14**, 988 (1976).
88. M. Kondo, H. Aoyagi, T. Kato and N. Izumiya, *Bull. Chem. Soc. Jpn.*, **39**, 2234 (1966).
89. S. Matsuura, M. Waki and N. Izumiya, *ibid.*, **45**, 863 (1972).
90. L. G. Snezhkova, E. N. Shepel, I. D. Ryabova, A. I. Miroshnikov, V. T. Ivanov and Yu. A. Ovchinnikov, *Bioorg. Khim.*, **1**, 347 (1975).
91. H. Aoyagi and N. Izumiya, *Bull. Chem. Soc. Jpn.*, **39**, 1747 (1966).
92. S. Matsuura, M. Waki, S. Makisumi and N. Izumiya, *ibid.*, **43**, 1197 (1970).
93. S. Sōfuku, I. Muramatsu, K. Okada and A. Hagitani, *ibid.*, **48**, 2888 (1975).
94. S. Sōfuku, A. Yoshida, H. Baba and I. Muramatsu, *ibid.*, **50**, 2143 (1977).
95. G. W. Kenner and J. M. Turner, *Chem. Ind.* (*London*), **1955**, 602.

96. G. W. Kenner, P. J. Thomson and J. M. Turner, *J. Chem. Soc.*, **1958**, 4148.
97. P. M. Hardy, G. W. Kenner and R. C. Sheppard, *Tetrahedron*, **19**, 95 (1963).
98. H. Brockmann and V. Graef, *Naturwissenschaften*, **49**, 540 (1962).
99. K. Isono and R. W. Curtius, *Phytochemistry*, **3**, 277 (1964).
100. J. C. Sheehan and W. L. Richardson, *J. Am. Chem. Soc.*, **76**, 6329 (1954).
101. H. Brockmann, H. Tummes and F. -A.v. Metzsch, *Naturwissenschaften*, **41**, 37 (1954).
102. P. W. G. Smith, *J. Chem. Soc.*, **1957**, 3985.
103. J. C. Sheeden, M. Goodman and W. L. Richardson, *J. Am. Chem. Soc.*, **77**, 6391 (1955).
104. H. Brockmann and M. Springorum, *Naturwissenschaften*, **49**, 514 (1962).
105. J. A. Reader and P. W. G. Smith, *J. Chem. Soc.*, **1965**, 3479.
106. M. Kondo and N. Izumiya, *Bull. Chem. Soc. Jpn.*, **40**, 1975 (1967).
107. M. Kondo and N. Izumiya, *ibid.*, **43**, 1850 (1970).
108. M. Waki, O. Abe, R. Okawa, T. Kato, S. Makisumi and N. Izumiya, *ibid.*, **40**, 2904 (1967).
109. O. Abe and N. Izumiya, *ibid.*, **43**, 1202 (1970).
110. R. Nagata, M. Waki, M. Kondo, H. Aoyagi, T. Kato, S. Makisumi and N. Izumiya, *ibid.*, **40**, 963 (1967).
111. S. Lee, R. Ohkawa and N. Izumiya, *ibid.*, **44**, 158 (1971).
112. H. Aoyagi, T. Kato, M. Waki, O. Abe, R. Okawa, S. Makisumi and N. Izumiya, *ibid.*, **42**, 782 (1969).
113. M. Waki and N. Izumiya, *Tetrahedron Lett.*, **1968**, 3083.
114. M. Waki and N. Izumiya, *Bull. Chem. Soc. Jpn.*, **41**, 1909 (1968).
115. R. Schwyzer and P. Sieber, *Helv. Chim. Acta*, **43**, 1910 (1960).
116. M. Kawai and U. Nagai, *Peptide Chemistry* (in Japanese) (ed. H. Yajima), p. 130, Protein Research Foundation (1975).
117. H. Takiguchi, H. Nishikawa and N. Izumiya, *Fukuoka Univ. Sci. Rep.*, **3**, 31 (1974).
118. S. Sōfuku, *Bull. Chem. Soc. Jpn.*, **46**, 968 (1973).
119. S. Matsuura and N. Izumiya, *Experientia*, **28**, 1402 (1972).
120. S. Matsuura, M. Waki, T. Kato and N. Izumiya, *Bull. Chem. Soc. Jpn.*, **46**, 977 (1973).
121. H. Sugano, H. Abe, M. Miyoshi, T. Kato and N. Izumiya, *Experientia*, **29**, 1488 (1973).
122. H. Sugano, H. Abe, M. Miyoshi, T. Kato and N. Izumiya, *Bull. Chem. Soc. Jpn.*, **47**, 698 (1974).
123. H. Abe, K. Sato, T. Kato and N. Izumiya, *ibid.*, **49**, 3113 (1976).
124. H. Nishikawa, H. Takiguchi and N. Izumiya, *Fukuoka Univ. Sci. Rep.*, **7**, 103 (1977).
125. L. Yu. Sklyarov and I. V. Shashkova, *Zh. Obshch. Khim.*, **39**, 2778 (1969).
126. H. Takiguchi, H. Nishikawa, S. Ando and N. Izumiya, *Bull. Chem. Soc. Jpn.*, **51**, 297 (1978).
127. H. Takiguchi, H. Nishikawa and N. Izumiya, *Fukuoka Univ. Sci. Rep.*, **4**, 107 (1974).

128. S. Lee, H. Aoyagi, Y. Shimohigashi, N. Izumiya, T. Ueno and H. Fukami, *Tetrahedron Lett.*, **1976**, 843.
129. J. Halstrøm, T. Christensen and K. Brunfeldt, *Z. Physiol. Chem.*, **357**, 999 (1976).
130. H. Takiguchi, H. Nishikawa and N. Izumiya, *Fukuoka Univ. Si. Rep.*, **5**, 139 (1975).
131. K. Nonaka, K. Sato, S. Terada, T. Kato and N. Izumiya, *Bull. Chem. Soc. Jpn.*, **51**, 2127 (1978).
132. A. A. Akhrem, A. S. Tumash, N. A. Poddubnaya and A. V. Otchnik, *Vestsi Akad. Navuk B. SSR, Ser. Khim. Navuk,* **1976**, 99.
133. K. Sato, K. Ueda, M. Kondo, H. Aoyagi and N. Izumiya, *Bull. Chem. Soc. Jpn.*, **51**, 1830 (1978).
134. O. Abe, Y. Utsumi and N. Izumiya, *ibid.*, **50**, 2341 (1977).
135. M. Rothe and F. Eisenbeiss, *Angew. Chem. Int. Ed. Engl.*, **7**, 883 (1968).
136. M. Rothe and F. Eisenbeiss, *Z. Naturforsch.*, **21b**, 814 (1966).
137. J. Halstrøm and H. Klostermeyer, *Justus Liebigs Ann. Chem.*, **715**, 208 (1968).
138. M. Fridkin, A. Patchornik and E. Katchalski, *J. Am. Chem. Soc.*, **87**, 4646 (1965).
139. M. Fridkin, A. Patchornik and E. Katchalski, *Biochemistry*, **11**, 466 (1972).
140. R. Kalir, A. Warshawsky, M. Fridkin and A. Patchornik, *Eur. J. Biochem.*, **59**, 55 (1975).
141. E. Flanigan and G. R. Marshall, *Tetrahedron Lett.*, **1970**, 2403.
142. H. Klostermeyer, J. Halstrøm, P. Kusch, J. Föhles and W. Lunkenheimer, *Peptides* (ed. H. C. Beyerman *et al.*), p. 113, North-Holland Pub. Co. (1967).
143. A. A. Kiryushkin, Yu. A. Ovchinnikov, I. V. Kozhevnikova and M. M. Shemyakin, *ibid.*, p. 100.
144. Yu. A. Ovchinnikov, V. T. Ivanov and M. M. Shemyakin, *ibid.*, p. 173.
145. A. I. Miroshnikov, L. G. Snezhkova, S. V. Sychev, I. I. Chervin, L. B. Senyavina, V. T. Ivanov and Yu. A. Ovchinnikov, *Bioorg. Khim.*, **3**, 180 (1977).
146. M. M. Shemyakin, Yu. A. Ovchinnikov, V. T. Ivanov and I. D. Ryabova, *Experientia*, **23**, 326 (1967).
147. M. M. Shemyakin, Yu. A. Ovchinnikov and V. T. Ivanov, *Angew. Chem. Int. Ed. Engl.*, **8**, 492 (1969).
148. J. R. McDermott and N. L. Benoiton, *Chemistry and Biology of Peptides* (ed. J. Meienhofer), p. 369, Ann Arbor Science Pub. Inc. (1972).
149. J. R. McDermott and N. L. Benoiton, *Can. J. Chem.*, **51**, 2562 (1973).
150. V. M. Stepanov, A. B. Silaev and A. N. Polin, *Antibiotiki*, **3**, 49 (1958).
151. N. Nishino, H. Aoyagi, T. Kato and N. Izumiya, *J. Biochem.* (*Tokyo*), **82**, 901 (1977).
152. V. T. Ivanov, A. I. Miroshnikov, L. G. Snezhkova, Yu. A. Ovchinnikov, A. V. Kulikov and G. I. Likhtenshtein, *Khim. Prir. Soedin.*, **9**, 91 (1973).
153. K. Okamoto, H. Abe, K. Kuromizu and N. Izumiya, *Mem. Fac. Sci. Kyushu Univ. Ser. C*, **9**, 78 (1974).
154. M. Kawai and U. Nagai, *Chem. Lett.*, **1977**, 1397.

155. M. Ohno and N. Izumiya, *J. Am. Chem. Soc.*, **88**, 376 (1966).
156. M. Ohno, T. Kato, S. Makisumi and N. Izumiya, *Bull. Chem. Soc. Jpn.*, **39**, 1738 (1966).
157. K. Kuromizu and N. Izumiya, *Experientia*, **26**, 587 (1970).
158. K. Kuromizu and N. Izumiya, *Bull. Chem. Soc. Jpn.*, **43**, 2199 (1970).
159. K. Kuromizu and N. Izumiya, *Tetrahedron Lett.*, **1970**, 1471.
160. K. Kuromizu and N. Izumiya, *Bull Chem. Soc. Jpn.*, **43**, 2944 (1970).
161. N. Mitsuyasu and N. Izumiya, *Experientia*, **26**, 476 (1970).
162. N. Mitsuyasu, S. Matsuura, M. Waki, M. Ohno, S. Makisumi and N. Izumiya, *Bull. Chem. Soc. Jpn.*, **43**, 1829 (1970).
163. K. Okamoto, M. Waki, T. Kato and N. Izumiya, *Mem. Fac. Sci. Kyushu Univ. Ser. C*, **8**, 297 (1973).
164. K. Okamoto, K. Nonaka and N. Izumiya, *Bull. Chem. Soc. Jpn.*, **50**, 231 (1977)
165. K. Okamoto, H. Yonezawa and N. Izumiya, *J. Chromatogr.*, **92**, 147 (1974).
166. S. Laiken, M. Printz and L. C. Craig, *J. Biol. Chem.*, **244**, 4454 (1969).
167. M. A. Ruttenberg, T. P. King and L. C. Craig, *Biochemistry*, **5**, 2857 (1966).
168. J. I. Harris and T. S. Work, *Biochem. J.*, **46**, 196 (1950).
169. J. I. Harris and T. S. Work, *ibid.*, **46**, 582 (1950).
170. B. F. Erlanger, H. Sachs and E. Brand, *J. Am. Chem. Soc.*, **76**, 1806 (1954).
171. B. F. Erlanger and L. Goode, *Nature*, **174**, 840 (1954).
172. B. F. Erlanger and L. Goode, *Science*, **131**, 669 (1960).
173. L. W. Pollard, N. V. Bhagavan and J. B. Hall, *Biochemistry*, **7**, 1153 (1968).
174. S. Makisumi and N. Izumiya, *Tetrahedron Lett.*, **1970**, 227.
175. S. Makisumi, S. Matsuura, M. Waki and N. Izumiya, *Bull. Chem. Soc. Jpn.*, **44**, 210 (1971).
176. S. Makisumi, M. Waki and N. Izumiya, *Mem. Fac. Sci. Kyushu Univ. Ser. C*, **8**, 1 (1972).
177. S. Makisumi, M. Waki and N. Izumiya, *Bull. Chem. Soc. Jpn.*, **44**, 143 (1971).
178. T. Nakamiya, H. Mizuno, T. Meguro, H. Ryono and K. Takinami, *J. Ferment. Technol.*, **54**, 369 (1976).
179. M. A. Ruttenberg, T. P. King and L. C. Craig, *Biochemistry*, **3**, 758 (1964).
180. A. I. Miroshnikov, St. Iordanov, E. N. Shepel, V. T. Ivanov and Yu. A. Ovchinnikov, *Peptides 1972* (ed. H. Hanson and H-D. Jakubke), p. 341, North-Holland Pub. Co. (1973).
181. B. R. Srinivasa and L. K. Ramachandran, *Indian J. Biochem. Biophys.*, **14**, 35 (1977).
182. E. Katchalski, L. Bichowsky-Slomnitzki and E. Valcani, *Biochem. J.*, **55**, 671 (1953).
183. L. Bichowsky-Slomnicki, A. Berger, J. Kurtz and E. Katchalski, *Arch. Biochem. Biophys.*, **65**, 400 (1956).
184. E. Katchalski, A. Berger, L. Bichowsky-Slomnicki and J. Kurtz, *Nature*, **176**, 118 (1955).
185. N. Izumiya, T. Kato, Y. Fujita, M. Ohno and M. Kondo, *Bull. Chem. Soc. Jpn.*, **37**, 1809 (1964)

186. K. Ikeda, K. Kuromizu, K. Okamoto and N. Izumiya, *Chem. Pharm. Bull.*, **20**, 1849 (1972).
187. R. Weindling, *Phytopathology*, **31**, 991 (1941).
188. M. R. Bell, J. R. Johnson, B. S. Wildi and R. B. Woodward, *J. Am. Chem. Soc.*, **80**, 1001 (1958).
189. A. F. Beecham, J. Fridrichsons and A. McL. Mathieson, *Tetrahedron Lett.*, **1966**, 3131.
190. E. Francis, R. Rahman, S. Safe and A.T aylor, *J. Chem. Soc. Perkin Trans. 1*, **1972**, 470.
191. R. Nagarajan, L. L. Huckstep, D. H. Lively, D. C. DeLong, M. M. Marsh and N. Neuss, *J. Am. Chem. Soc.*, **90**, 2980 (1968).
192. R. Nagarajan, J. L. Occolowitz, N. Neuss and S. M. Nash, *J. Chem. Soc. Chem. Commun.*, **1969**, 359.
193. G. S. Rosenfeld, L. I. Rostovtseva, V. M. Baikima, D. M. Trakhtemberg and A. S. Khokhlov, *Antibiotiki*, **8**, 201 (1963).
194. A. S. Khokhlov and G. B. Lokshin, *Tetrahedron Lett.*, **1963**, 1881.
195. T. Kato, M. Kondo, M. Ohno and N. Izumiya, *Bull. Chem. Soc. Jpn.*, **38**, 1202 (1965).
196. O. Abe, K. Kuromizu, M. Kondo and N. Izumiya, *ibid.*, **43**, 914 (1970).
197. H. Takiguchi and N. Izumiya, *ibid.*, **47**, 221 (1974).
198. R. Schwyzer and P. Sieber, *Helv. Chim. Acta*, **41**, 1582 (1958).
199. M. Kondo and N. Izumiya, *Ann. Mtg. Chem. Soc. Japan* (in Japanese), 1972. *Abstracts*, III, p. 1290.
200. A. L. Zhuze, G. A. Kogan, N. A. Krit, T. M. Andronova, M. P. Filatova, L. B. Senyavina, E. A. Meshcheryakova, I. D. Ryabova, G. A. Ravdel and L. A. Shchukina, *Mol. Biol. (Moscow)*, **8**, 84 (1974).
201. Yu. A. Ovchinnikov and V. T. Ivanov, *Tetrahedron*, **31**, 2117 (1975).
202. Yu. A. Ovchinnikov, *Peptides 1972* (ed. H. Hanson and H-D. Jakubke), p. 3, North-Holland Pub. Co. (1973).
203. I. Muramatsu, S. Sōfuku and A. Hagitani, *J. Antibiot.*, **25**, 189 (1972).
204. S. Hiramatsu, *J. Osaka City Med. Cent.* (in Japanese), **8**, 2017 (1959).
205. A. B. Silaev, Humboldt-Symp. Grundfragen Biol., Berlin, p. 156 (1960); *Chem. Abstr.*, **59**, 6635 (1963).
206. A. S. Kaprel'yants, V. V. Nikiforov, A. I. Miroshnikov, L. G. Snezhkova, V. A. Eremin and D. N. Ostrovskii, *Biokhimiya*, **42**, 329 (1977).
207. T. Uehara, *J. Osaka City Med. Cent.* (in Japanese), **8**, 1499 (1959).
208. A. B. Silaev and V. M. Stepanov, *Dokl. Akad. Nauk SSSR*, **112**, 297 (1957).
209. Y. Hashimoto, *J. Osaka City Med. Cent.* (in Japanese), **9**, 4019 (1960).
210. V. M. Stepanov and A. B. Silaev, *Zh. Obshch. Khim.*, **31**, 3811 (1961).
211. T. Uehara, *J. Osaka City Med. Cent.* (in Japanese), **8**, 1489 (1959).
212. T. Yukawa, *ibid.*, **6**, 285 (1957).
213. V. M. Stepanov and A. B. Silaev, *Zh. Obshch. Khim.*, **31**, 3804 (1961).
214. I. P. Kuranova and N. V. Smirnova, *Izv. Akad. Nauk SSSR, Otd. Khim. Nauk*, **1963**, 1148.
215. W. Patton, E. Jellum, D. Nitecki, W. Pereira and B. Halpern, *Aust. J. Chem.*, **22**, 2709 (1969).

216. A. A. Akhrem, A. S. Tumash and N. A. Poddubnaya, *Vestsi Akad. Navuk B. SSR, Ser. Khim. Navuk*, **1976**, 95.
217. R. Chandrasekaran, A. V. Lakshminarayanan, U. V. Pandya and G. N. Ramachandran, *Biochim. Biophys. Acta*, **303**, 14 (1973).
218. C. M. Venkatachalam, *Biopolymers*, **6**, 1425 (1968).
219. T. Kato and N. Izumiya, *Biochim. Biophys. Acta*, **493**, 235 (1977).
220. M. Yukioka, Y. Saito and S. Otani, *J. Biochem.* (*Tokyo*), **60**, 295 (1966).
221. N. Takahashi, T. Kato and N. Izumiya, *Mem. Fac. Sci. Kyushu Univ. Ser. C*, **9**, 71 (1974).
222. L. S. Izotova, D. I. Gorodetsky, V. V. Yanonis, L. A. Baratova, L. P. Belyanova, E. A. Timokhina, A. Ya. Strongin and V. M. Stepanov, *Bioorg. Khim.*, **4**, 397 (1978).
223. N. Nishino, H. Aoyagi, T. Kato and N. Izumiya, *Experientia*, **31**, 410 (1975).
224. S. Terada, K. Sato, T. Kato and N. Izumiya, *FEBS Lett.*, **90**, 89 (1978).
225. P. C. Montgomery, J. H. Rockey and A. R. Williamson, *Proc. Natl. Acad. Sci. U.S.A.*, **69**, 228 (1972).
226. J. T. Edsall, P. J. Flory, J. C. Kendrew, A. M. Liquori, G. Némethy, G. N. Ramachandran and H. A. Scheraga, *J. Biol. Chem.*, **241**, 1004 (1966).
227. IUPAC-IUB Commission on Biochemical Nomenclature, *ibid.*, **245**, 6489 (1970).
228. V. F. Bystrov, V. T. Ivanov, S. L. Portnova, T. A. Balashova and Yu. A. Ovchinnikov, *Tetrahedron*, **29**, 873 (1973).
229. D. C. Hodgkin and B. M. Oughton, *Biochem. J.*, **65**, 752 (1957).
230. N. B. Abbott and E. J. Ambrose, *Proc. R. Soc. London, Ser. A*, **219**, 17 (1953).
231. C. Camilletti, P. De Santis and R. Rizzo, *J. Chem. Soc. Chem. Commun.*, **1970**, 1073.
232. W. Moffitt and J. T. Yang, *Proc. Natl. Acad. Sci. U.S.A.*, **42**, 596 (1956).
233. F. Quadrifoglio and D. W. Urry, *Biochem. Biophys. Res. Commun.*, **29**, 785 (1967).
234. A. M. Liquori, P. De Santis, A. L. Kovacs and L. Mazzarella, *Nature*, **211**, 1039 (1966).
235. D. W. Urry, A. L. Ruiter, B. C. Starcher and T. A. Hinners, *Antimicrob. Agents Chemother.*, **1968**, 87.
236. D. Balasubramanian, *J. Am. Chem. Soc.*, **89**, 5445 (1967).
237. A. Stern, W. A. Gibbons and L. C. Craig, *Proc. Natl. Acad. Sci. U.S.A.*, **61**, 734 (1968).
238. R. W. Woody, *Peptides Polypeptides and Proteins* (ed. E. R. Blout *et al.*), p. 338, John Wiley & Sons (1974).
239. T. Kato, M. Waki, S. Matsuura and N. Izumiya, *J. Biochem.* (*Tokyo*), **68**, 751 (1970).
240. A. M. Liquori and F. Conti, *Nature*, **217**, 635 (1968).
241. F. Conti, *ibid.*, **221** 777 (1969).
242. E. M. Bradbury, B. G. Carpenter, C. Crane-Robinson and H. W. E. Rattle, *ibid.*, **220**, 69 (1968).

243. R. Schwyzer and U. Ludescher, Biochemistry, **7**, 2519 (1968).
244. M. Ohnishi and D. W. Urry, *Biochem. Biophys. Res. Commun.*, **36**, 194 (1969).
245. T. P. Pitner and D. W. Urry, *J. Am. Chem. Soc.*, **94**, 1399 (1972).
246. Yu. A. Ovchinnikov, V. T. Ivanov, V. F. Bystrov, A. I. Miroshnikov, E. N. Shepel, N. D. Abdullaev, E. S. Efremov and L. B. Senyavina, *Biochem. Biophys. Res. Commun.*, **39**, 217 (1970).
247. W. A. Gibbons, G. Némethy, A. Stern and L. C. Craig, *Proc. Natl. Acad. Sci. U.S.A.*, **67**, 239 (1970).
248. N. G. Kumar, N. Izumiya, M. Miyoshi, H. Sugano and D. W. Urry, *J. Am Chem. Soc.*, **97**, 4105 (1975).
249. N. G. Kumar, N. Izumiya, M. Miyoshi, H. Sugano and D. W. Urry, *Biochemistry*, **14**, 2197 (1975).
250. D. J. Patel and A. E. Tonelli, *Biopolymers*, **15**, 1623 (1976).
251. T. Kato, *Kagaku to Seibutsu* (in Japanese), **14**, 206 (1976).
252. W. A. Gibbons, H. Alms, R. S. Bockman and H. R. Wyssbrod, *Biochemistry*, **11**, 1721 (1972).
253. I. D. Rae, E. R. Stimson and H. A. Scheraga, *Biochem. Biophys. Res. Commun.*, **77**, 225 (1977).
254. W. A. Gibbons, D. Crepaux, J. Delayre, J.-J. Dunand, G. Hajdukovic and H. R. Wyssbrod, *Peptides: Chemistry, Structure and Biology*, (ed. R. Walter and J. Meienhofer), p. 127, Ann Arbor Science Pub. Inc. (1975).
255. W. A. Gibbons, J. A. Sogn, A. Stern, L. C. Craig and L. F. Johnson, *Nature*, **227**, 840 (1970).
256. J. A. Sogn, L. C. Craig and W. A. Gibbons, *J. Am. Chem. Soc.*, **96**, 3306 (1974).
257. F. A. Bovey, *Chemistry and Biology of Peptides* (ed. J. Meienhofer), p. 3, Ann Arbor Science Pub. Inc. (1972).
258. A. Allerhand and R. A. Komoroski, *J. Am., Chem. Soc.*, **95**, 8228 (1973).
259. G. E. Hawkes, E. W. Randall and C. H. Bradley, *Nature*, **257**, 767 (1975).
260. M. A. Khaled, D. W. Urry, H. Sugano, M. Miyoshi and N. Izumiya, *Biochemistry*, **17**, 2490 (1978).
261. R. T. Ingwall, C. Gilon and M. Goodman, *J. Am. Chem. Soc.*, **97**, 4356 (1975).
262. L. C. Craig, *Science*, **144**, 1093 (1964).
263. S. L. Laiken, M. P. Printz and L. C. Craig, *Biochemistry*, **8**, 519 (1969).
264. D. T. Warner, *Nature*, **190**, 120 (1961).
265. P. De Santis and A. M. Liquori, *Biopolymers*, **10**, 699 (1971).
266. G. Vanderkooi, S. J. Leach, G. Némethy, R. A. Scott and H. A. Scheraga, *Biochemistry*, **5**, 2991 (1966).
267. R. A. Scott, G. Vanderkooi, R. W. Tuttle, P. M. Shames and H. A. Scheraga, *Proc. Natl. Acad. Sci. U.S.A.*, **58**, 2204 (1967)
268. F. A. Momany, G. Vanderkooi, R. W. Tuttle and H. A. Scheraga, *Biochemistry*, **8**, 744 (1969).
269. M. Digert, N. Go and H. A. Scheraga, *Macromolecules*, **8**, 750 (1975).
270. E. S. Pysh, *Science*, **167**, 290 (1970).

271. A. Stern, W. A. Gibbons and L. C. Craig, *J. Am. Chem. Soc.*, **91**, 2794 (1969).
272. W. A. Gibbons, C. F. Beyer, J. Dadok, R. F. Sprecher and H. R. Wyssbrod, *Biochemistry*, **14**, 420 (1975).
273. U. Ludescher and R. Schwyzer, *Helv. Chem. Acta*, **54**, 1637 (1971).
274. U. Ludescher and R. Schwyzer, *ibid.*, **55**, 2052 (1972).
275. M. M. Shemyakin, Yu. A. Ovchinnikov, V. T. Ivanov and A. V. Evstratov, *Nature*, **213**, 412 (1967).
276. M. Kawai and U. Nagai, *Peptide Chemistry*, (ed. T. Nakajima), p. 101, Protein Research Foundation (1977).
277. T. Higashijima, M. Tasumi, T. Miyazawa, M. Kawai and U. Nagai, *Peptide Chemistry* (ed. T. Shiba), p. 97, Protein Research Foundation (1978).
278. I. Uemura, C. J. Chern, T. Saito and Y. Hashimoto, *J. Osaka City Med. Cent.* (in Japanese), **9**, 3625 (1960).
279. Y. Miki, *ibid.*, **9**, 4005 (1960).
280. T. Hiramatsu, *ibid.*, **10**, 267 (1961).
281. W. Pache, D. Chapman and R. Hillaby, *Biochim. Biophys. Acta*, **255**, 358 (1972).
282. M. Koike, K. Iida and T. Matsuo, *J. Bacteriol.*, **97**, 448 (1969).
283. K. S. Rosenthal, P. E. Swanson and D. R. Storm, *Biochemistry*, **15**, 5783 (1976).
284. F. M. Harold and J. R. Baarda, *J. Bacteriol.*, **94**, 53 (1967).
285. M. C. Goodall, *Biochim. Biophys. Acta*, **219**, 471 (1970).
286. W. R. Veatch, E. T. Fossel and E. R. Blout, *Biochemistry*, **13**, 5249 (1974).
287. K. Jayaraman and R. Kannan, *Biochem. Biophys. Res. Commun.*, **48**, 1235 (1972).
288. T. Yoshida, H. Weissbach and E. Katz, *Arch. Biochem. Biophys.*, **114**, 252 (1966).
289. N. Sarkar and H. Paulus, *Nature New Biol.*, **239**, 228 (1972).
290. P. K. Mukherjee and H. Paulus, *Proc. Natl. Acad. Sci. U.S.A.*, **74**, 780 (1977).
291. N. Sarkar, D. Langley and H. Paulus, *ibid.*, **74**, 1478 (1977).
292. B. Schazschneider, H. Ristow and H. Kleinkauf, *Nature*, **249**, 757 (1974).
293. H. Ristow, B. Schazschneider, J. Vater and H. Kleinkauf, *Biochim. Biophys. Acta*, **414**, 1 (1975).
294. H. Ristow, B. Schazschneider, K. Bauer and H. Kleinkauf, *ibid.*, **390**, 246 (1975).
295. H. J. Ristow, B. Schazschneider and H. Kleinkauf, *Biochem. Biophys. Res. Commun.*, **63**, 1085 (1975).
296. D. Neubert and A. L. Lehninger, *Biochim. Biophys. Acta*, **62**, 556 (1962).
297. M. Kambe, Y. Imae and K. Kurahashi, *J. Biochem.* (*Tokyo*), **75**, 481 (1974).
298. G. M. Church, J. L. Sussman and S. -H. Kim, *Proc. Natl. Acad. Sci. U.S.A.*, **74**, 1458 (1977).
299. K. Kurahashi, *Ann. Rev. Biochem.*, **43**, 445 (1974).
300. F. Lipmann, *Accounts Chem. Res.*, **6**, 361 (1973).

301. C. Gilhuus-Moe, T. Kristensen, J. E. Bredesen, T. L. Zimmer and S. G. Laland, *FEBS Lett.*, **7**, 287 (1970).
302. S. Tomino and K. Kurahashi, *Biochem. Biophys. Res. Commun.*, **17**, 288 (1964).
303. T. Tsuji, *J. Osaka City Med. Cent.* (in Japanese), **15**, 1 (1966).
304. H. Holm, L. O. Frøholm and S. G. Laland, *Biochim. Biophys. Acta*, **115**, 361 (1966).
305. S. Otani, T. Yamanoi, Y. Saito and S. Otani, *Biochem. Biophys. Res. Commun.*, **25**, 590 (1966).
306. S. Tomino, M. Yamada, H. Itoh and K. Kurahashi, *Biochemistry*, **6**, 2552 (1967).
307. H. Kleinkauf, W. Gevers and F. Lipmann, *Proc. Natl. Acad, Sci. U.S.A.*, **60**, 269 (1968).
308. M. Yamada and K. Kurahashi, *J. Biochem. (Tokyo)*, **66**, 529 (1969).
309. M. Yamada and K. Kurahashi, *ibid.*, **63**, 59 (1968).
310. T. Ljones, O. Vaage, T. L. Zimmer, L. O. Frøholm and S. G. Laland, *FEBS Lett.*, **1**, 339 (1968).
311. J. E. Bredesen, T. L. Zimmer and S. G. Laland, *ibid.*, **3**, 169 (1969).
312. W. Gevers, H. Kleinkauf and F. Lipmann, *Proc. Natl. Acad. Sci. U.S.A.*, **63**, 1335 (1969).
313. R. Roskoski, Jr., G. Ryan, H. Kleinkauf, W. Gevers and F. Lipmann, *Arch. Biochem. Biophys.*, **143**, 485 (1971).
314. H. Kleinkauf, R. Roskoski, Jr. and F. Lipmann, *Proc. Natl. Acad. Sci. U.S.A.*, **68**, 2069 (1971).
315. E. Stoll, Ø. Frøyshov, H. Holm, T. L. Zimmer and S. G. Laland, *FEBS Lett.*, **11**, 348 (1970).
316. H. Saxhalm, T. L. Zimmer and S. G. Laland, *Eur. J. Biochem.*, **30**, 138 (1972).
317. S. G. Laland, Ø. Frøyshov, C. Gilhuus-Moe and T. L. Zimmer, *Nature New Biol.*, **239**, 43 (1972).
318. R. Roskoski, Jr., W. Gevers, H. Kleinkauf and F. Lipmann, *Biochemistry*, **9**, 4839 (1970).
319. R. Roskoski, Jr., H. Kleinkauf, W. Gevers and F. Lipmann, *ibid.*, **9**, 4846 (1970).
320. M. Kambe, Y. Sakamoto and K. Kurahashi, *J. Biochem. (Tokyo)*, **69**, 1131 (1971).
321. K. Fujikawa, Y. Sakamoto and K. Kurahashi, *ibid.*, **69**, 869 (1971).
322. K. Kurahashi, M. Yamada, K. Mori, K. Fujikawa, M. Kambe, Y. Imae, E. Sato, H. Takahashi and Y. Sakamoto, *Cold Spring Harbor Symp. Quant. Biol.*, **34**, 815 (1969).
323. H. Kleinkauf, W. Gevers, R. Roskoski, Jr. and F. Lipmann, *Biochem. Biophys. Res. Commun.*, **41**, 1218 (1970).
324. B. Mach and E. L. Tatum, *Proc. Natl. Acad. Sci. U.S.A.*, **52**, 876 (1964).
325. R. E. Winnick and T. Winnick, *Biochim. Biophys. Acta*, **53**, 461 (1961).
326. F. Lipmann, W. Gevers, H. Kleinkauf and R. Roskoski, Jr., *Adv. Enzymol.*, **35**, 1 (1971).

327. K. Okuda, G. C. Edwards and T. Winnick, *J. Bacteriol.*, **85**, 329 (1963).
328. H. Ishihara and K. Shimura, *Biologically Active Peptides Produced by Microorganisms* (in Japanese) (ed. H. Umezawa, T. Shiba and T. Takita), p. 11, Kyoritsu Shuppan (1976).
329. E. Katz and A. L. Demain, *Bacteriol. Rev.*, **41**, 449 (1977).
330. N. D. Fulton, K. Bollenbacher and G. E. Templeton, *Phytopathology*, **55**, 49 (1965).
331. M. Koncewicz, P. Mathiaparanam, T. F. Uchytil, L. Sparapans, J. Tam, D. H. Rich and R. D. Durbin, *Biochem. Biophys. Res. Commun.*, **53**, 653 (1973).
332. W. L. Mayer, L. F. Kuyper, D. W. Phelpo and A. W. Cordes, *J. Chem. Soc. Chem. Commun.*, **1974**, 339.
333. W. L. Mayer, G. E. Templeton, C. I. Grable, R. Jones, L. F. Kuyper, R. B. Lewis, C. W. Sigel and S. H. Woodhead, *J. Am. Chem. Soc.*, **97**, 3802 (1975).
334. A. Hirota, A. Suzuki, H. Suzuki and S. Tamura, *Agric. Biol. Chem.*, **37**, 643 (1973).
335. A. Hirota, A. Suzuki and S. Tamura, *ibid.*, **37**, 1185 (1973).
336. A. Hirota, A. Suzuki, K. Aizawa and S. Tamura, *ibid.*, **37**, 955 (1973).
337. A. Closse and R. Huguenin, *Helv. Chim. Acta*, **57**, 533 (1974).
338. Y. Shimohigashi and N. Izumiya, *J. Synth. Org. Chem.* (in Japanese), **36**, 1023 (1978).
339. K. Sawamura, *Bull. Hort. Res. Stan. Jpn., Ser. C* (in Japanese), **4**, 43 (1966).
340. T. Okuno, Y. Ishita, A. Sugawara, Y. Sakan and T. Matsumoto, *Proc. 18th Symp. Chemistry of Natural Products, Kyoto* (in Japanese), p. 361, 1974.
341. T. Okuno, Y. Ishita, S. Nakayama, K. Sawai, T. Fujita and K. Sawamura, *Ann. Phytopath. Soc. Jpn.*, **40**, 375 (1974).
342. T. Okuno, Y. Ishita, K. Sawai and T. Matsumoto, *Chem. Lett.*, **1974**, 635.
343. T. Ueno, Y. Hayashi, T. Nakashima, H. Fukami, S. Nishimura and A. Sekiguchi, *Ann. Phytopath. Soc. Jpn.* (in Japanese), **40**, 174 (1974).
344. T. Ueno, T. Hayashi, T. Nakashima, H. Fukami, S. Nishimura, K. Kohmoto and A. Sekiguchi, *Phytopathology*, **65**, 82 (1975).
345. T. Ueno, T. Nakashima, Y. Hayashi and H. Fukami, *Proc. 18th Symp. Chemistry of Natural Products, Kyoto* (in Japanese), p. 369, 1974.
346. T. Ueno, T. Nakashima, Y. Hayashi and H. Fukami, *Agric. Biol. Chem.*, **39**, 1115 (1975).
347. T. Ueno, T. Nakashima, Y. Hayashi and H. Fukami, *ibid.*, **39**, 2081 (1975).
348. K. Kohmoto, I. D. Khan, Y. Renbutsu, T. Taniguchi and S. Nishimura, *Physiol. Plant Path.*, **8**, 141 (1976).
349. T. Okuno, Y. Ishita, A. Sugawara, Y. Mori, K. Sawai and T. Matsumoto, *Tetrahedron Lett.*, **1975**, 335.
350. E. Gross and J. L. Morell, *J. Am. Chem. Soc.*, **89**, 2791 (1967).
351. E. Gross, J. L. Morell and L. C. Craig, *Proc. Natl. Acad. Sci. U.S.A.*, **62**, 952 (1969).
352. T. Ueno, T. Nakashima, M. Uemoto, H. Fukami, S. Lee and N. Izumiya, *Biomed. Mass Spectromet.*, **4**, 134 (1977).

353. N. Izumiya and A. Nagamatsu, *Bull. Chem. Soc. Jpn.*, **25**, 265 (1952)
354. N. Izumiya, S. Matsuura and K. Kuromizu, *J. Synth. Org. Chem.* (in Japanese), **29**, 1032 (1971).
355. S. Lee, N. Izumiya, A. Suzuki and S. Tamura, *Tetrahedron Lett.*, **1975**, 883.
356. S. Lee and N. Izumiya, *Int. J. Peptide Protein Res.*, **10**, 206 (1977).
357. H. Schulz, *Chem. Ber.*, **99**, 3425 (1966).
358. E. Gross, K. Noda and B. Nisula, *Angew. Chem. Int. Ed. Engl.*, **12**, 664 (1973).
359. D. H. Rich and P. Mathiaparanam, *Tetrahedron Lett.*, **1974**, 4037.
360. S. Lee, H. Aoyagi, Y. Shimohigashi, N. Izumiya, T. Ueno and H. Fukami, *Proc. 19th Symp. Chemistry of Natural Products, Hiroshima* (in Japanese), p. 342, 1975.
361. Y. Shimohigashi, S. Lee, H. Aoyagi, T. Kato and N. Izumiya, *Int. J. Peptide Protein Res.*, **10**, 197 (1977).
362. I. Photaki, *J. Am. Chem. Soc.*, **85**, 1123 (1963).
363. Y. Shimohigashi, S. Lee, H. Aoyagi, T. Kato and N. Izumiya, *Int. J. Peptide Protein Res.*, **10**, 323 (1977).
364. Y. Shimohigashi, S. Lee and N. Izumiya, *Bull. Chem. Soc. Jpn.*, **49**, 3280 (1976).
365. D. H. Rich, P. Mathiaparanam, J. A. Grant and P. Phatnagar, *Peptides: Chemistry, Structure and Biology* (ed. R. Walter and J. Meienhofer), p. 943, Ann Arbor Science (1975).
366. Y. Shimohigashi, S. Lee, T. Kato, N. Izumiya, T. Ueno and H. Fukami, *Agric. Biol. Chem.*, **41**, 1533 (1977).
367. Y. Shimohigashi, S. Lee, T. Kato and N. Izumiya, *Bull. Chem. Soc. Jpn.*, **51**, 584 (1978).
368. T. Kato, S. Lee, Y. Shimohigashi, H. Aoyagi and N. Izumiya, *Mem. Fac. Sci., Kyushu Univ., Ser. C*, **10**, 179 (1978).
369. Y. Shimohigashi, S. Lee, T. Kato, N. Izumiya, T. Ueno and H. Fukami, *Chem. Lett.*, **1977**, 1411.
370. O. Wintersteiner and M. Moore, *J. Am. Chem. Soc.*, **65**, 1507 (1943).
371. H. Ako, C. H. Ryan and R. J. Foster, *Biochem. Biophys. Res. Commun.*, **46**, 1639 (1972).
372. E. M. Chamberlin, E. W. Tristram, T. Utne and J. M. Chemerda, *J. Org. Chem.*, **25**, 295 (1960).
373. G. G. Hazen and D. W. Resonburg, *ibid.*, **29**, 1930 (1964).
374. J. L. Flippen and I. L. Karle, *Biopolymers*, **15**, 1081 (1976).
375. A. Yasutake, S. Lee, H. Aoyagi, T. Kato and N. Izumiya, *Chem. Lett.*, **1978**, 875.
376. J. Konnert and I. L. Karle, *J. Am. Chem. Soc.*, **91**, 4888 (1969).
377. N. Go and H. A. Scheraga, *Macromolecules*, **3**, 178 (1970).
378. T. Higashijima, M. Tasumi, T Miyazawa, Y. Shimohigashi, T. Kato and N. Izumiya, *Peptide Chemistry* (ed. N. Izumiya), p. 121, Protein Research Foundation (1979).
379. A. Sekiguchi, T. Ueno, H. Fukami, Y. Hayashi, T. Nakashima and S. Nishimura, *Ann. Phytopath. Soc. Jpn.* (in Japanese), **40**, 149 (1974).

380. T. Ueno, *J. Agric. Chem. Soc. Jpn.* (in Japanese), **50**, R141 (1976).
381. K. Kohmoto, T. Taniguchi and S. Nishimura, *Ann. Phytopath. Soc. Jpn.*, **43**, 65 (1977).
382. S. Tsuchiya, Y. Yoshida and T. Haniuda, *Bull. Hort. Res. Stan. Jpn., Ser. C* (in Japanese), **5**, 9 (1967).
383. I. Kozaki, *ibid., Ser. A* (in Japanese), **12**, 17 (1973).
384. A. Sekiguchi, T. Ueno, Y. Hayashi, T. Nakashima, H. Fukami and S. Nishimura, *Ann. Phytopath. Soc. Jpn.* (in Japanese), **41**, 262 (1975).
385. S. Nishimura, K. Kohmoto and H. Otani, *Rev. Plant Protection Res.*, **7**, 21 (1974).
386. H. Otani, S. Nishimura, K. Kohmoto, K. Yano and T. Seto, *Ann. Phytopath. Soc. Jpn.*, **41**, 467 (1975).
387. S. E. Hull, R. Karlsson, P. Main, M. M. Woolfson and E. J. Dodson, *Nature*, **275**, 206 (1978).
388. M. J. E. Sternberg and J. M. Thornton, *ibid.*, **271**, 15 (1978).

Index

A

B

T

W, X, Z